THE WATERSHED

Also by Annabel Soutar

Novembre (2000)

2000 Questions (2002)

Import/Export (2008)

Sexy béton (2010)

*Seeds** (2012)

Fredy (2016)

*Published by Talonbooks

THE WATERSHED

A DOCUMENTARY PLAY

ANNABEL SOUTAR

TALONBOOKS

Talonbooks
278 East First Avenue, Vancouver, British Columbia, Canada V5T 1A6
www.talonbooks.com

First printing: 2016

Typeset in Frutiger Serif
Printed and bound in Canada on 100% post-consumer recycled paper

Cover design by Typesmith
Cover image © Kevin Dinkel (http://kevindinkel.com)
Interior design by Jenn Murray

Talonbooks acknowledges the financial support of the Canada Council for the Arts, the Government of Canada through the Canada Book Fund, and the Province of British Columbia through the British Columbia Arts Council and the Book Publishing Tax Credit.

LIBRARY AND ARCHIVES CANADA CATALOGUING IN PUBLICATION

Soutar, Annabel, 1971–, author
The watershed : a documentary play / Annabel Soutar.

Issued in print and electronic formats.
ISBN 978-0-88922-988-4 (paperback). – ISBN 978-0-88922-989-1 (epub).
– ISBN 978-1-77201-048-0 (kindle). – ISBN 978-1-77201-049-7 (pdf)

I. Title.

PS8637.O94W37 2016 C812'.6 C2016-901728-1
C2016-901729-X

For Ian Alexander Soutar, 1936–2016,
who taught me about the ultimate form of eloquence: action

A NOTE ON THE TEXT

The text of *The Watershed* is made up of verbatim testimony from interviews conducted by the playwright, as well as speech transcripts, House of Commons debate transcripts, film transcripts, press releases, and media articles collected between 2013 and 2015. A certain degree of artistic licence was taken with verbatim script derived from interview material with family members.

Chris Abraham's production was performed by eight actors who each played a number of roles, except for the actress playing only the character of Annabel. Abraham's role breakdown, however, does not constitute an imperative cast distribution for future productions.

Syntactical and grammatical "errors" in the script reflect the idiosyncrasies of the characters' verbatim language and should therefore be respected in performance.

PRODUCTION HISTORY

The Watershed was first produced as a co-production between Montreal's Porte Parole Productions and Crow's Theatre of Toronto, and was performed July 7 to 19, 2015, at the Berkeley Street Theatre in Toronto as part of and commissioned by Panamania, the arts festival of the Toronto 2015 Pan Am and Parapan Am Games, and also developed in part through The Collaborations, an initiative of the National Arts Centre of Canada, English Theatre. The play received dramaturgical workshop support from Emma Tibaldo and Playwright's Workshop Montreal. Cast and crew included the following:

Bruce Dinsmore	Don Shipley, Dr. Hank Venema, James, Stephen Harper, Dave Gillis, David McGuinty (LIB MP), Cameron (CBC TV reporter), Jian Ghomeshi, Film Narrator, Male Sushi Diner, Oil Sands Tour Pilot
Alex Ivanovici	Alex, Rambling Dan Frechette, Dr. Paul Frost, Kennedy Stewart (NDP MP), Dr. Mike Paterson
Tanja Jacobs	Janice Dean (Fox News reporter), Minister Shelly Glover, Hazel, Maude Barlow, Granny, Michelle Rempel (CPC MP), Acting Speaker of the House (Barry Devolin), Adèle Hurley, Oil Man, Kathleen Wynne, Carol Off
Tara Nicodemo	Mayor Rob Ford, Fort Garry Hotel Waitress, Susanne McCrea, Caroline, Eloise Mitchell, Claudia Barilà, Angela (CBC TV producer), Greg Rickford (CPC MP), Liisa (Chris Abraham's wife), Female Sushi Diner

Ngozi Paul	Beatrice, Chris Abraham, Dr. Maggie Xenopoulos, Anonymous Liberal MP, Assistant to Senator Seidman
Eric Peterson	Fox News Anchor, Elliot Levine (the plumber), Ontario Minister Michael Chan (LIB MPP), Dr. David Schindler, Grandpa, Minister Gary Goodyear (CPC MP), John (CBC TV cameraman), Minister Joe Oliver (CPC MP), First Nations Elder, Sushi Waiter
Amelia Sargisson	Ella, Diane Orihel, Lauren Liu (NDP MP)
Kristen Thomson	Annabel

Director	Chris Abraham
Set and Costume Designer	Julie Fox
Sound Designer	Thomas Ryder Payne
Lighting Designer	Kimberly Purtell
Video Projection Designer	Denyse Karn
Assistant to the Director	Andrew Kushnir
Assistant to the Playwright	Elle Thoni
Stage Manager	Merissa Tordjman

Le partage des eaux, a French-language version of the play, was performed at Montreal's Usine C, November 17 to 21, 2015.

CHARACTERS

(*in order of appearance*)

ANNABEL: a Montreal-based documentary theatre creator, early forties

ALEX: playwright's husband and father of Ella and Beatrice, early forties

ELLA: Annabel and Alex's daughter, age ten

BEATRICE: Annabel and Alex's daughter, age eight

NEWS ANCHOR: male, any age

JANICE DEAN: reporter, Fox News Weather Center, mid-forties

ELLIOT: plumber, sixties

SHELLY GLOVER: then federal minister of Canadian heritage, late forties

MICHAEL CHAN: then Ontario minister of tourism, culture, and sport, sixties

ROB FORD: then mayor of Toronto, forties

DON SHIPLEY: Pan Am Games creative director for arts and culture, festivals, sixties

DIANE ORIHEL: Ph.D. candidate in ecology at the University of Alberta and director of the Coalition to Save ELA, thirties

CHRIS ABRAHAM: artistic director of Crow's Theatre in Toronto, Annabel's co-production partner, forties

HAZEL: Chris's daughter, age seven, friend of Ella and Beatrice

DR. HENRY (HANK) VENEMA: vice-president, business development, International Institute for Sustainable Development, forties

DR. DAVID SCHINDLER: University of Alberta ecology professor and freshwater scientist, sixties

MAUDE BARLOW: national chairperson for the Council of Canadians, sixties

SUSANNE McCREA: executive director, Boreal Forest Network, forties

RAMBLING DAN FRECHETTE: singer-songwriter, thirties

CAROLINE: Annabel's friend, forties

GRANDPA: Ian, Annabel's father, seventies

GRANNY: Helgi, Annabel's mother, seventies

JAMES: Annabel's brother, uncle to Ella and Beatrice, forties

ABBY: James's wife, aunt to Ella and Beatrice, forties

STEPHEN HARPER: then prime minister of Canada, fifties

ELOISE MITCHELL: then Department of Fisheries and Oceans communications manager, forties

DAVE GILLIS: then Department of Fisheries and Oceans assistant deputy minister, late sixties

DR. MAGGIE XENOPOULOS: professor of aquatic science, Trent University, thirties

DR. PAUL FROST: Maggie's husband, limnologist and professor of aquatic science, Trent University, thirties

KENNEDY STEWART: then NDP MP for Burnaby-Douglas, fifties

DAVID MCGUINTY: then Liberal MP for Ottawa South, fifties

MICHELLE REMPEL: then CPC MP for Calgary Nose Hill, parliamentary secretary to the minister of the environment, thirties

LAURIN LIU: then NDP MP for Rivière-des-Mille-Îles, twenties

GARY GOODYEAR: then CPC MP for Cambridge, fifties

ACTING SPEAKER OF THE HOUSE (BARRY DEVOLIN): then CPC MP for Haliburton-Kawartha Lakes-Brock, fifties

CAMERON: CBC TV reporter, mid-forties

ANGELA: CBC TV producer, early thirties

JOHN: CBC TV cameraman, sixties

ADÈLE HURLEY: director, Program on Water Issues, University of Toronto Munk School of Global Affairs, fifties

JOE OLIVER: then federal minister of natural resources, CPC MP for Eglinton-Lawrence, early seventies

CLAUDIA BARILÀ: Annabel's friend and spouse of Cirque du Soleil and One Drop Foundation co-founder Guy Laliberté, thirties

DR. MIKE PATERSON: then ELA senior research scientist, late fifties

ASSISTANT TO SENATOR SEIDMAN: female, thirties

OIL MAN: an executive vice-president (sustainability) for a large oil sands company, forties

JIAN GHOMESHI: then host of CBC Radio show *Q*, forties

KATHLEEN WYNNE: Liberal premier of Ontario, sixties

CAROL OFF: host of CBC Radio show *As It Happens*, sixties

GREG RICKFORD: then CPC MP for Kenora, fifties

LIISA: Hazel's mother and Chris's wife, forties

FIRST NATIONS ELDER: seventies

FILM NARRATOR: thirties

MALE SUSHI DINER: thirties

FEMALE SUSHI DINER: thirties

OIL SANDS TOUR PILOT: male, thirties

Other characters, played by various actors: two waitresses, workman, Canadian Press reporter

Approximate run time: 3 hours, including one 20-minute intermission

ACT 1

SCENE 1

ALEX, ANNABEL, ELLA, and BEATRICE are sitting on the front stoop of their house in the Plateau Mont-Royal neighbourhood of Montreal. ELLA and BEATRICE are eating ice cream cones. ALEX and ANNABEL are looking at their smartphones. It is an unseasonably hot, late October afternoon in 2012.

ANNABEL

I called the plumber. (*pause*) Do you think it's the best way to start the research?

ALEX

I think ... water is a big subject. (*pause*) You gotta start somewhere.

ANNABEL

I know.

ALEX

So when's he coming?

ANNABEL

Sunday. It's his only day off.

ALEX

It's our only day off.

ANNABEL

Darling, our life is a day off.

Breaking news from the Fox News Weather Center, October 25, 2012.

NEWS ANCHOR

And a Fox News Alert happening now on Hurricane Sandy. That storm already slamming Cuba, now we have brand new tracking which shows where this massive system might be heading next. And it is not good news for those who live in the northeastern United States and eastern Canada. Janice Dean has the latest from the Fox News Extreme Weather Center. What's goin' on, J.D.?

JANICE DEAN

If this track comes true it is the worst-case scenario, John. Right now this storm looks like it's going to be headed for millions of people. If you live across the mid-Atlantic, up to Maine and eastern Canada, you need to be paying very, very close attention. The *worst-case scenario* happening right now, people. I am not just standing here making this stuff up. We are going to see torrential rain and flooding inland, high winds – significant power outages that could last for weeks for millions of people. John, our mouths have just dropped at the latest tracking system that shows this storm heading right for us!

The sound of the storm arrives in the family's universe. ALEX, ANNABEL, ELLA, and BEATRICE sit transfixed by the oncoming storm. It appears to slam directly into their house before they can even react.

SCENE 2

Two days later. Sound of very loud drying machines. The house is occupied by four industrial-sized dryer fans blowing air up into huge holes in the home's walls and ceilings. ELLIOT is ringing the front doorbell. ANNABEL sees him through the window.

ANNABEL

Oh shit ... shoot ... the plumber! Alex! Elliot's here!

ALEX

Who?

ANNABEL

The plumber!

ALEX

What?!

ANNABEL opens the front door. We can hear her dialogue with ELLIOT now that she's outside. Sound of a dog barking furiously inside the house.

ANNABEL

Elliot, I'm so sorry.

ELLIOT

Is this a bad time?

ANNABEL

No, it's actually perfect. You won't believe what happened. Come in.

ELLIOT enters the noisy household.

ELLIOT

Wow!

ANNABEL

Can I get you a glass of water or something?

ELLIOT

No, I'm good. Jesus!

ANNABEL turns off the dryer fans one by one.

Offstage sound of Katy Perry's "Extraterrestrial."

ANNABEL

Alex! Girls! Elliot is here!

ALEX

(*offstage*) Okay, girls! Turn it off!

ELLIOT

Lemme guess ... your pipes couldn't handle the storm?

ANNABEL

Yes, and the speedway got disconnected somehow. But it's okay – the insurance is covering it. We have to redo all the bathrooms. Girls!

ELLIOT

Jeez. Sorry about the piano.

ANNABEL

Yeah. (*pause*) Elliot, I just want to say, um, you know we would have hired your company to do the work but –

ELLIOT

You don't even have to say it ... I know. The insurance guys always have their own plumbing teams.

ELLIOT glances around the house imagining all the money he could have made.

ALEX

(*offstage*) I SAID TURN IT OFF!!

Music stops abruptly.

ALEX enters, followed by ELLA.

ALEX

Elliot, how are you?

ELLIOT

Always a pleasure, Mr. Ivanovici. (*to ELLA*) Hello, I'm Elliot.

ELLA

Hi.

ELLIOT

What's your name?

ELLA

Um, Ella.

ELLIOT

Ella. Like Elliot.

ALEX

It's very similar. You wanna shake Elliot's hand?

ELLA hesitates.

ELLIOT

How old are you, Ella?

ELLA

Ten.

ELLIOT

Give me five!

ELLA slaps ELLIOT's hand half-heartedly.
BEATRICE enters.

BEATRICE

Mama, I'm hungry. Can I have a Fruit Roll-Up?

ELLA

Oh, me too!

ELLIOT

Hi!

BEATRICE

Hi.

ANNABEL

Bibi, would you like to meet Elliot first?

ELLIOT

What's your name?

Pause.

ANNABEL

Can you say hello to Elliot?!

ELLIOT

Hi! I'm Elliot. How are you?

BEATRICE

I'm Beatrice.

ELLIOT

Hi, Beatrice! How old are you?

BEATRICE

Eight.

Pause.

ELLIOT

Okay then, uh, whenever you're ready ...

ANNABEL passes BEATRICE and ELLA two Fruit Roll-Ups – extremely long, sticky fake-fruit snacks that the girls will unspool and play with during the interview.

ANNABEL

So the reason I asked Elliot to come to the house is because I thought, as a point of departure for a play about water, that we need to become more *in tune* with how water enters, circulates, and leaves our house – and Beatrice, you can suck on that, but maybe a little more discreetly please – because we don't even really know how water gets into our house, right?

ELLA

Yes, we do. By pipes.

ANNABEL

Okay, but –

ELLIOT

That's correct, Ella!

ANNABEL

But do you know how the pipes get in here?

BEATRICE

Yeah, from underground.

ELLIOT

That's right. You girls are good!

BEATRICE

And the pipes go all around the house and all over the country!

ELLIOT

They do! They do! So if the pipes are all hidden underground, what happens when there's a pipe that leaks underneath the street?

ELLA

Uh ...

ELLIOT

How do we ever *find* the leaky part of the pipe?

BEATRICE gestures to answer, but ELLIOT cuts her off.

ELLIOT

So I'll tell you! Down underground with the pipes there are *pumps*. They provide the *pressure* inside of the pipes. (*pumping his arm up and down*) Now what happens is that the pump is pumping away and the pressure doesn't build up. So they know that there's a major leak somewhere!

BEATRICE

I don't get it –

ANNABEL

Shh!

ELLIOT

They calculate that about 50 percent of our drinking water that they pump outta the river is lost into the ground!

BEATRICE has stopped paying attention and is playing with her Fruit Roll-Up.

ANNABEL

That's terrible. Did you hear that, Beatrice?

BEATRICE

Yeah.

ELLA

No! *She* didn't, but I did.

BEATRICE

Yes, I did!

ANNABEL

Okay, so what did Elliot just say?

BEATRICE

Uhhh.

BEATRICE giggles.

ELLA

What he said is that –

ANNABEL

What *Elliot* said.

ELLA

What Elliot said is that there, um, in the pumps there is, like, like, some water leaking.

ANNABEL

Yeah. How much? How much of it?

ELLA

Well, 50 percent of ... of it ... of the water that they pimp ... pump from the river is going, um, well, is, like, leaking out of the pumps because there's maybe sometimes tiny, tiny little holes the water can get through. Beatrice, I'm explaining this for *you* and you're not even listening!

BEATRICE

Leave me alone, Ella!

ALEX

OKAY, THAT'S ENOUGH!

SCENE 3

ELLIOT is gone. Family debriefs.

ANNABEL

This is a privilege, you know? That we are taking you on one of our projects.

BEATRICE

Well, it doesn't feel like a privilege.

ALEX

What does it feel like?

BEATRICE

It feels like a boring.

ALEX

Like "a" boring?

BEATRICE

No, it's *boring*.

ALEX

Why is it boring to you?

BEATRICE

Because –

ELLA

To me it wasn't boring it –

BEATRICE

It –

ELLA

It's just a little –

ANNABEL

Let Beatrice answer the question.

BEATRICE

Because, like, it's a bit boring because, like, for ... for ... really ... for adults, they love talking but kids they love playing, so just standing there and listening –

ANNABEL

Okay.

BEATRICE

– listening to an adult that you don't even understand ... It's like a story that you don't even understand and you're in the middle of it.

ANNABEL

Did you not understand the story Elliot was telling us? That we spend tens of millions of dollars treating and delivering fresh water into our homes each year and *60 percent* of that –

ELLA

50 percent.

ANNABEL

– 50 percent of that just leaks into the ground before we can get it! I mean ...!

BEATRICE

Well ... but Elliot used, like, fancy words so I didn't understand anything.

ALEX

And if he's used a fancy word you didn't understand, Bibi, why didn't you ask him what it meant?

BEATRICE

Because Mummy always says not to interrupt people.

Pause.

ALEX

So this might come as a surprise to you girls but do you know why Mummy wants you to help her research her new play?

BEATRICE

Because she doesn't know how else to start writing it.

ALEX

No. Because she *wants* to be interrupted.

BEATRICE

No, she doesn't.

ALEX

Yes, she does. She wants your help. This play is going to be about water and the future. Your future. That's why we're recording all of our family conversations about water. Mummy's going to use your words in the play.

ELLA

Is that true, Mummy?

ANNABEL

Yes. Some of them.

BEATRICE/ELLA

Cool.

ALEX

It *is* cool.

ANNABEL

But I'm going to say one thing now that I want everyone to understand, Ella too.

ELLA

(*petulant*) What?

ANNABEL

That there is never any problem with interrupting to *ask a question.*

SCENE 4

A press conference to announce cultural programming for the Toronto 2015 Pan Am and Parapan Am Games. Present onstage are then federal Minister of Canadian Heritage SHELLY GLOVER, then Ontario Minister of Tourism, Culture, and Sport MICHAEL CHAN, then Mayor of Toronto ROB FORD, and head of programming for the games DON SHIPLEY.

DON SHIPLEY

One of the big questions of our time is: What is the future of water? And what can we do, especially in Canada, to protect that future? Minister Glover –

Sound of applause.

Minister Chan –

Sound of applause.

Mayor Ford –

Sound of less applause.

Ladies and gentlemen, the organizers of Panamania hope to make the Toronto 2015 Pan Am and Parapan Am Games as much about arts and culture as about sport. And specifically, using art to get people thinking and talking about these big questions. It is our hope that the artists we have commissioned – including Robert Lepage, Veronica Tennant, Rick Miller ... and Annabel Soutar ...

ANNABEL reluctantly steps forward.

DON SHIPLEY

... with a total of one-and-a-half million dollars for our AquaCulture programming, will show the magnificence of water beyond sports – its cultural possibilities. At the end of the day we are sufficiently blessed here in Canada in terms of abundance of water, but we are facing serious issues.

Transition to ANNABEL at home. ALEX joins ANNABEL after putting BEATRICE in the bathtub.

ALEX

How was the press conference?

ANNABEL

I felt like an imposter.

ALEX

Well, the girls are asking me how pipes separate the hot water from the cold water, so I guess something sunk in. (*pause*) Have you ever heard of something called the Experimental Lakes Area?

ANNABEL

No.

ALEX

I hadn't either, which is weird because it's one of the most important freshwater research sites in the world. And guess where it's located?

ANNABEL

Israel?

ALEX

Canada. Check this out.

Sound of Death of Evidence rally.

ANNABEL and ALEX watch video footage on a laptop showing the Death of Evidence rally on Parliament Hill in Ottawa in July 2012.

ALEX

This summer a bunch of scientists marched on Parliament Hill because the federal government announced funding cuts to this place.

ANNABEL

Who's that woman in black?

ALEX

Diane Orihel.

SCENE 5

At home. ANNABEL conducts a Skype interview with DIANE ORIHEL, director of the Coalition to Save ELA, while BEATRICE takes a bath.

DIANE

During the late 1960s, Lake Erie was being choked with blue-green algae. And the government at the time needed answers for how to fix those problems, because there was a debate in the scientific literature about which nutrient in sewage is actually causing the algae to bloom. Some people were saying it's carbon, some people were saying it's nitrogen, some people were saying it's phosphorous. And the government at the time realized that they couldn't figure it out in Lake Erie. They needed to do experiments on small lakes and figure out that answer somewhere else.

ANNABEL

Why can't you figure it out in Lake Erie?

DIANE

(*chuckling*) Because it's a really, really, really big lake. And there are many different factors that are affecting the health of the lake.

ANNABEL

Of course, you need to isolate.

DIANE

Yes, and the reason the ELA is so vital is because there is nowhere else in the world where there are fifty-eight lakes that have been set aside and dedicated for research on fresh water.

ANNABEL

So why would our government defund the ELA right now?

DIANE

Many scientists believe this is a very strategic decision. They wanted the research to stop because it's inconvenient to their agenda.

ANNABEL

But they must be smart enough to know they can't block knowledge?

DIANE

Well ... but that's what they're actually trying to do. This is systemic, Annabel. This government is gutting environmental legislation, firing world-class scientists. In the last budget, they eliminated nearly three thousand environmental assessments. And there's going to be nothing left to protect fresh water once this government is done. So it's time for us to take more risks as scientists and speak out.

ANNABEL

Are enough people doing that? Your colleagues at ELA, other scientists?

DIANE

Well, this is one of the things that has really astounded me, actually – that people are afraid to speak out, not only within government where scientists are explicitly warned not to speak. But some academic scientists are afraid to get blacklisted if they speak publicly. Even students. This blows my mind. You would think they would be the most vocal. But some of them have even been scared to sign our petitions because they think it will affect their ability to get a job as government scientists in the future.

ANNABEL

Wow.

DIANE

Yeah, I wasn't prepared for that actually.

ANNABEL

But you have had a lot of traction with the media, right?

DIANE

Yeah, and with the public. More than 25,000 Canadians have signed the Save ELA petition. I've been working with the opposition MPs, like, they have been tabling these petitions almost every day in the House of Commons. These ELA petitions have been tabled I think more than eighty times now. But still no response from the government. It's like talking to a brick wall.

ANNABEL

So what is it going to take?

DIANE

I wish I knew the answer, because it seems like whatever it is that we do ... it just ... (*pause*) We've had several open letters published in high-profile journals and papers like the *Globe and Mail*. We've sent letters to the prime minister, to Minister Ashfield, to Minister Kent ... we've requested meetings. All we get back are form letters.

ANNABEL

This sounds like a full-time job.

DIANE laughs.

DIANE

Yeah.

ANNABEL

How are you managing to do this while continuing your own research?

DIANE

Uh, well, I've had to put my studies on hold because, well, this consumes every waking minute. My advisers are getting pretty impatient.

ANNABEL

Well, they must understand that what you are doing now will affect not only your research but the research of many other people for years to come, so ...?

DIANE

Yeah, they do see that and they are supportive. But I guess they're also trying to look out for me. Since I'm not.

DIANE laughs but almost cries.

ANNABEL

This is my daughter by the way – Beatrice. This is Diane.

DIANE

Hi, Beatrice.

BEATRICE

Hi.

DIANE

It's really nice to meet you.

ANNABEL

My daughters are actually helping me do interviews ... but ... Beatrice was in the bath this time.

BEATRICE

Is she going to do the water project with us?

ANNABEL

Well, you know what? Diane is already on a water project.

BEATRICE

She is?

ANNABEL

She's a water *scientist.*

BEATRICE

Woooah.

ANNABEL

She is trying to understand what happens when different pollutants go into lakes, how it affects the fish and the plants, and all that stuff.

BEATRICE

(*shy*) Are you learning that in school?

ANNABEL

You have to speak up, Beatrice.

BEATRICE

Are you learning that in school?

DIANE

Yeah, I'm a student – just like you.

SCENE 6

ANNABEL is on a Skype call with CHRIS ABRAHAM, her co-production partner from Crow's Theatre in Toronto. CHRIS's daughter, HAZEL, is with him. ELLA and ALEX in the house somewhere.

CHRIS

So you've decided?

ANNABEL

Not ... entirely.

CHRIS

Why not?

ANNABEL

It's just tricky because it's a decision I have to make so quickly, like, almost now.

CHRIS

Okay. Why?

ANNABEL

Because after speaking with Diane, like, stuff is starting to go down. Like, she's organizing a big Save ELA rally in Winnipeg next week ...

CHRIS

Uh-huh?

ANNABEL

... and on March 31, which is in, like, three months, the federal funding cut is going into effect.

HAZEL

Dad, can I talk with Ella and Beatrice?

CHRIS

Not now, Hazel.

ELLA

Hazel!

HAZEL and ELLA squeal.

ANNABEL

Just wait a second, Ella. You can talk with Hazel after I'm finished with Chris.

ALEX

Ella, you're not finished your math.

ANNABEL

(*to CHRIS*) So, like, the Coalition to Save ELA is racing against that deadline to mobilize people. Maude Barlow's going to be there. This other important freshwater scientist is doing a talk. So I can hit them all in one go.

CHRIS

So?

ANNABEL

So if I want to have a hope in hell of covering this story properly, I have to fly to Winnipeg on Tuesday – alone.

CHRIS

Well, sounds to me like your decision is made. (*pause*) Annabel, it's not an either-or thing. We just can't afford for you to take everybody everywhere.

ANNABEL

I know.

CHRIS

So I'll talk to Monica. But I think we should disperse some of the Pan Am cash for your plane ticket.

Passing his laptop to HAZEL.

Okay, Hazel, she's all yours.

HAZEL and ELLA take over the Skype call.

SCENE 7

The Fort Garry Hotel, Winnipeg, November 2012. ANNABEL runs into DIANE in the lobby. DR. HANK VENEMA, International Institute for Sustainable Development (IISD), is standing nearby at the lobby bar.

ANNABEL

Diane!

DIANE

You made it.

ANNABEL

Of course.

DIANE

I'm so glad. Oh God, what time is it? I have to go set up.

ANNABEL

Can I help you?

DIANE

No, it's cool. Get a drink. I'll see you up there.

DIANE exits. ANNABEL approaches HANK at the bar.

ANNABEL

Hello. Are you here for the rally?

HANK

As a matter of fact, I am.

ANNABEL

Me too. I'm Annabel.

HANK

How are you, Annabel? I'm Hank.

ANNABEL

Hi, Hank.

HANK

Where are you from?

ANNABEL

I'm from Montreal.

HANK

Are you media?

ANNABEL

Um, I'm a playwright.

HANK

You're a playwright. Well, that's media. (*chuckling*) What exactly are you writing about?

ANNABEL

I'm writing about water in Canada.

ANNABEL hands HANK a brochure describing her project. The WAITRESS serves HANK a drink and starts to walk away.

ANNABEL

(*to WAITRESS*) Can I have a glass of water, please?

The WAITRESS ignores her.

HANK

(*looking at brochure*) Where are your kids?

ANNABEL

What?

HANK

Says in your blurb here you're going to research this story with your kids.

ANNABEL

Oh yeah. We couldn't afford to all fly to Winnipeg.

HANK

The Pan Am Games, huh?

ANNABEL

Yup.

HANK

So how sanitized does your story have to be to play at the Pan Am Games?

Pause.

ANNABEL

Sanitized? Not at all, I don't think.

HANK

Good. Then you're in the right place.

ANNABEL

What is ... your role in all this?

DR. DAVID SCHINDLER walks across hotel lobby. HANK sees him.

HANK

(*handing ANNABEL a business card*) We'll talk about that some other time. Nice to meet you.

HANK leaves, following in the direction of SCHINDLER.

ANNABEL looks at card.

SCENE 8

Later that evening in a ballroom at the Fort Garry Hotel. MAUDE BARLOW speaks to the ELA rally audience, among them DIANE, HANK, SCHINDLER, ANNABEL, and SUSANNE. The onstage audience applauds and heckles during MAUDE's speech.

MAUDE BARLOW

I want to step back here for a minute and remember that we are a world running out of clean water. We were all taught back in Grade 6 that this wasn't possible. That we have a certain amount of finite water and it goes around and around in a hydrologic cycle and it can't go anywhere. But what humans are doing is polluting, mismanaging, and most importantly displacing land-based water to places where we can no longer get at it. We are currently doubling groundwater takings every twenty years.

If the Great Lakes are being depleted as quickly as the underground water around the world, the Great Lakes could be bone-dry in eighty years. The Ogallala Aquifer in the United States that irrigates the breadbasket for the U.S. – it will be gone in our lifetime. That is a statement from the U.S. Department of Agriculture. We are literally losing the groundwater for future generations in one generation.

The World Bank recently announced that demand for fresh water is growing so much faster than supply, that by the year 2030 – which is not very far away – demand for water will outstrip supply by 40 percent.

Any country that is as blessed with water resources as we are here in Canada has a special responsibility to take care of that water for future generations. That is exactly what scientists at the ELA have been doing for more than four decades.

So why are its funds being cut right now? Well, if you're a government that wants to implement rampant exploitation of natural resources, that's what you do. You remove any obstacles to that exploitation. You go after a research site where evidence is gathered about the impact of industry on fresh water. You go after the independent scientist who knows how to read that evidence. And you try quietly to change the law without anyone noticing – by introducing a 450-page omnibus bill – Bill C-38 (supposedly a "budget bill") – with changes to environmental legislation buried deep within it that have nothing to do with the budget.

So we are here tonight to build resistance to these actions, to point out that these actions add up to a carefully crafted agenda – an assault on independent science – which is an assault on the foundation of democracy, because without evidence-based science we will all remain ignorant.

Well, I for one refuse to be ignorant and I for one refuse to be quiet.

Enthusiastic applause.

SUSANNE

Wow! Maude Barlow, ladies and gentlemen. (*pause*) Good evening, I'm Susanne McCrea, and I'm the executive director of the Boreal Forest Network, and we are here tonight to, uh, formally announce that we are the newest member of the Coalition to Save ELA!

Applause.

It's my pleasure to join forces with someone that I've only recently begun to work with and I'm extremely impressed with.

Um, this is the best kind of activist to me – somebody who, in the course of their regular life, comes across something that they believe is *wrong* and stands up for themselves. And stands up for their principles. And this is Diane Orihel.

Applause for DIANE.

SUSANNE

I'd also like to acknowledge how lucky we are to have Dr. David Schindler with us here tonight – winner of the first Stockholm Water Prize and a co-founder of the Experimental Lakes Area.

Applause for SCHINDLER, who smiles humbly from a corner of the room. HANK is standing next to him.

SUSANNE

We look forward to hearing from Dr. Schindler in a few moments. But before we do we are going to continue our evening with a live performance of "My Canada" by Rambling Dan Frechette.

RAMBLING DAN FRECHETTE performs "My Canada" while SCHINDLER and HANK chat.

SCHINDLER

Our economists were predicting exactly this three years ago –

HANK

What, you mean if the price gets chronically above eighty bucks per barrel?

SCHINDLER

Yep. And the other thing is the crazy escalation in costs. They've gotta raise several billion dollars upfront for their new projects and they won't turn a nickel for –

ANNABEL

Excuse me, would you two mind ...? I'm sorry, I just couldn't help overhearing ... can I ... record any part of this conversation?

HANK

Annabel is a playwright.

SCHINDLER

Oh.

HANK

You know, I would prefer you didn't, to be totally honest.

ANNABEL

Can I ask why not?

HANK

Uh yes, you can. Because, before you ... before you walked over here I confessed to Dr. Schindler ... (*to ANNABEL*) Have you met David Schindler?

ANNABEL

No, hello.

SCHINDLER

Hello.

ANNABEL

Do you have a card ... I could ...?

SCHINDLER

I'm sorry, I don't.

HANK

Anyway, I had confessed to Dr. Schindler that we are deliberately trying to keep a low profile tonight because we are –

ANNABEL

Sorry, "we" being?

ANNABEL points to HANK and SCHINDLER.

HANK

No ... we ... we're not–

SCHINDLER

Oh ... no.

HANK

–together ... in terms of ... So "we" (*gesturing the idea of being separate from SCHINDLER*) being *me*, and "we" (*gesturing the idea of an organization*) being the IISD.

ANNABEL

I see. So why is the IISD trying to keep a low profile?

HANK

Shh! (*lowering voice*) Because if the stars align and, you know, we play our cards right–and if manna falls from heaven–we may become the new operator for the ELA.

ANNABEL

The new ... operator?

HANK

However, if it becomes formal and public ... we lose any amount of leverage we have.

Applause after musical performance.

SCHINDLER

Excuse me. I've got to get up there now.

HANK

Catch up with you later, Dr. Schindler.

After the event. ANNABEL intercepts DIANE in the lobby.

DIANE

Sorry, I've ... I've been so busy.

ANNABEL

That's okay. You did a great job.

DIANE

Thanks.

ANNABEL

Can we talk now for a sec?

DIANE

Yeah. I'm heading upstairs if you want to ride the elevator with me.

DIANE and ANNABEL get into an elevator.

ANNABEL

I just spoke to that guy Hank Venema of the IS–?

DIANE

The IISD.

ANNABEL

Yeah. I'm confused. I didn't know it was an option to move the ELA outside of government.

DIANE

It's not an option in my opinion.

ANNABEL

So why is that guy at your event? He was very friendly with David Schindler too.

DIANE

Uh, well –

ANNABEL

Is that too controversial for you to speak about?

DIANE

The IISD does not want to be identified as a potential operator. And, uh, the government does not want to speak about IISD being a potential operator.

ANNABEL

Why not?

DIANE

And uh, so, I have to respect both their wishes.

ANNABEL

Why is it a secret?

DIANE

I can't really –

DIANE and ANNABEL exit elevator.

DIANE

I'm sorry. I'm just not allowed to talk about it on the record.

ANNABEL

But you want the ELA to stay in the public sector, right?

DIANE

Oh yeah. I'm sorry, I have to meet with some of my coalition partners now.

ANNABEL

Oh, okay.

DIANE

Uh, I'm sorry but they won't be open to having a recorder in the room.

ANNABEL

I understand. No, you have to have your meeting.

DIANE

Sorry, I know you came all this way –

ANNABEL

No, that's okay. Can we ... maybe Skype early next week so I can ask you some follow-up questions?

DIANE

Um, yeah.

SCENE 9

ANNABEL arrives home to find ALEX reading Harry Potter *to ELLA and BEATRICE before bed.*

ALEX

"And Dumbledore awarded fifty points to Gryffindor House!"

BEATRICE/ELLA

Mummy!

ELLA

Was Diane nice?

ANNABEL

Really nice. She's an angel. A tough angel.

BEATRICE

Is she going to save the ELA?

ANNABEL

I think so. But not all by herself. She may need other people to help her.

ELLA

Why couldn't *we* go to Winnipeg?

ANNABEL

Let's not start that again. Next time, I promise. Good night.

ALEX

Take Charlotte with you.

ELLA and BEATRICE grudgingly go to bed.

ALEX

So? How'd it go?

ANNABEL

This whole thing is a little more complicated than I thought.

ALEX

Well, that's a good thing, right?

ANNABEL

Yeah.

SCENE 10

CAROLINE – a good friend of ANNABEL's – is over for dinner with ANNABEL and ALEX.

ELLA and BEATRICE are on the PlayStation.

CAROLINE

I know this may sound boring –

ALEX

Turn it down, girls!

CAROLINE

But frankly, that this ELA closure is some big sinister conspiracy to "muzzle science"? It's much more likely that someone in Harper's office charged some of his minions with: "Look at every discrete program that has an annual budget of ten million dollars and less. And by virtue of the fact that it's only ten million dollars or less, of what consequence can it really be to the national good?"

ANNABEL

Mm-hmm.

CAROLINE

I mean, I'm not defending Harper. But as someone who has managed a budget, when you have to cut, you take the low-hanging fruit first. (*pause*) That was delicious. I ate too fast.

ALEX

(*to ELLA and BEATRICE*) Okay, time to turn it off.

CAROLINE

So, what's the research task?

ANNABEL

I need you to find out some info about this organization.

She shows CAROLINE the IISD website on a laptop.

CAROLINE

IISD?

ANNABEL

International Institute for Sustainable Development.

Sound of a movie suddenly playing loudly.

ALEX

(*to ELLA and BEATRICE*) Off! Do it!

CAROLINE starts to surf around the IISD website. ELLA and BEATRICE eventually come closer to listen in on adult's conversation.

ANNABEL

So this is what I've been able to gather. ELA is currently a federally funded research site, run by DFO – the Department of Fisheries and Oceans. Last May DFO announces that ELA's federal funding is cut and Diane springs into action to mobilize resistance against that cut.

ALEX

She wants to make sure ELA remains federally funded.

ANNABEL

Exactly. But then I meet this guy at her event called Hank Venema –

CAROLINE

(*reading from website*) VP, Science and Innovation.

ANNABEL

Yes, who tells me that IISD may become the new *operator* of ELA but that he has to be totally hush-hush about the transfer. So, like, what's the deal? Why is this guy Hank at her rally when his purpose seems to be at odds with Diane's? And ... is the IISD some kind of oil-friendly special interest group dressed up as an environmental NGO?

CAROLINE shoots ANNABEL a skeptical look.

ANNABEL

I know, I know. But – first rule of research: Don't rule anything out.

CAROLINE

So what do you want to know that's not on their website?

ANNABEL

Their annual budget, their revenue sources, their capacity to take over an entire federal government program, their relationship to oil companies.

CAROLINE

So basically you want me to read their financial statements.

ANNABEL

And decode them for your flaky artist friend.

CAROLINE

Okay, I'm on it.

She gets up to leave.

ANNABEL

Thank you.

CAROLINE

Don't thank me 'til I've produced something. And can I be blunt about this ELA story? It can't be "Save the Fish," Annabel. It can't be that. It needs to be: This is the issue and you've got to immediately get to why it's relevant to humanity. And the only way to make the story relevant is to create the economic argument around it. (*pause*) Love what you're doing. My two cents.

CAROLINE exits. ANNABEL goes to her laptop.

SCENE 11

ANNABEL writes an email to DIANE. ALEX sits with ELLA and BEATRICE.

Dear Diane,
It was so inspiring meeting you in Winnipeg last week. I have tons of follow-up questions I'd love to ask you, if and when you have a moment. Especially about the IISD – if you can talk about that. Please let me know when we can Skype next.

BEATRICE
What's an economic argument, Daddy?

ALEX
It's an argument you use to convince people they should do something because it will make them money. Does that make sense?

Pause.

BEATRICE
Yeah.

DIANE enters and speaks in response to ANNABEL's earlier email message.

DIANE
(*automated email response*) Thank you kindly for your interest in the future of the Experimental Lakes Area. Please note that Diane Orihel has returned to her Ph.D. studies and is no longer serving as the director of the Coalition to Save ELA.

ANNABEL

What?

DIANE

(*automated email response*) For issues related to the Coalition to Save ELA, please email info@saveela.org.

ANNABEL

Shit.

ALEX

What's going on?

DIANE

(*automated email response*) Help us keep up the fight to save ELA.

ANNABEL

I have no idea.

DIANE

(*automated email response*) The health of Canada's fresh water depends on it! Thank you.

ANNABEL

Diane just ... up and left the Save ELA coalition.

SCENE 12

Christmas Eve, 2012, at the Soutar family cottage, Brome Lake, Quebec. GRANDPA, GRANNY, uncle JAMES, and aunt ABBY arrive to celebrate with ALEX, ANNABEL, ELLA, and BEATRICE. Lots of presents and champagne. They make a toast.

GRANDPA

Merry Christmas!

EVERYONE

Merry Christmas!

Afterwards, at the dinner table.

GRANDPA

Who wants ice cream?

ELLA

Oh! Me!

ALEX

I'll get it, Ian.

ANNABEL

So, Ella. Continue with what you were saying.

ELLA

So they don't want to, uh, give money for the ELA because, um, because they don't want the ELA to find out about all the pollution in our lakes.

GRANDPA

Who's "they"?

ELLA

Um, the government.

GRANDPA

The government is not a person.

ELLA

I mean, the president.

JAMES

You mean the prime minister.

ELLA

The prime minister.

JAMES

(*pouring wine*) This is gonna feel slightly more full-bodied than the 2007, Mum.

ELLA

Stephen Harper doesn't want anyone to find out about pollution in our lakes.

GRANDPA

Did Stephen Harper say that or did somebody else say that about him?

ANNABEL

Don't look at me.

ELLA

No! Stephen Harper doesn't want anybody to find out that oil is really, really extremely bad for the environment. Because then they'll stop buying oil for their cars. And then he's worried that, like, they'll stop buying cars.

GRANDPA

Well, he's worried that it'll hurt the *economy.*

ELLA

Yeah.

BEATRICE

Well, there's electric cars.

ELLA

Yeah but –

GRANNY

There *are* electric cars.

GRANDPA

But there's not many and they don't go that far.

ELLA

Yeah and –

GRANNY

There *aren't* that many. What's happening with everyone's grammar?

ELLA

– and Beatrice, electric cars have batteries which, if you throw them out, it's really, really bad for the environm–

BEATRICE sticks out her tongue at ELLA.

ELLA

Agh! Beatrice just stuck her tongue out at me!

BEATRICE

No, I didn't.

ELLA

Yes, you did! She did!

GRANNY

Now, girls.

ANNABEL

But Ella, maybe Grandpa has a different perspective. Do you want to ask him why *he* thinks Stephen Harper cut funds for the ELA?

ELLA

Why do you think, Grandpa?

GRANDPA

Well, girls, I think that the Canadian government is spending more money than they're taking in.

ELLA

Oh.

GRANDPA

And they're going more and more and more in debt, and when people go more and more and more in debt, it starts to have an impact on their standard of living and they start going backwards.

BEATRICE

Backwards?

GRANDPA

So the government is cutting back on all kinds of expenditures. (*to JAMES*) James, pour your mother some more vino. (*to ELLA and BEATRICE*) Now, you can argue that Mr. Harper is cutting back on the wrong things, but he's got very difficult choices to make. And he *has* to make those choices, otherwise, *you're* going to grow up and *you're* going to have to pay back all this debt – all the money that I'm enjoying and your parents are enjoying right now.

ELLA

How much money will we have to pay back?

GRANDPA

Well, if you look at the debt of Canada today, it is six hundred billion dollars.

ELLA and BEATRICE gasp.

GRANDPA

That's right. And it keeps going up and up and up every year. And somebody's going to have to pay it back one day. And it's going to be *you*.

BEATRICE

Billion ...?

GRANDPA

And so I think the government is wise to be trying to cut back on their expenditures.

Pause.

ELLA

I agree with you, Grandpa.

GRANDPA

Well, don't take my word for it, darling. But make sure you listen to a few different opinions. Not just Mummy's.

ALL drift off to go to bed except ANNABEL and GRANDPA, who finish their drinks after dinner.

ANNABEL

So why did your generation rack up so much debt then?

GRANDPA

We voted in too many Liberal governments. I did my bit to prevent it.

ANNABEL

I think the Conservatives are still spending quite a bit.

GRANDPA

Now that they have a majority, maybe they'll stop.

ANNABEL

Two million dollars a year on the best facility in the world to do research on fresh water? We can't afford that?

ANNABEL puts her recorder on the table.

GRANDPA

I guess this is for posterity now? (*pause*) Well, when you put it that way, of course we can afford it. But I mean, I don't know all the details of the federal balance sheet, do you?

ANNABEL

No.

GRANDPA

But I do imagine that they are spending a lot more than two million dollars a year on all of the environmental issues – I don't know the facts, Annabel.

ANNABEL

Dad, when an Order of Canada scientist who has won the Stockholm Water Prize writes a letter to the prime minister that says it's not a good idea to cut this, you would think that Stephen Harper would look up from his balance sheet for a second and respond.

GRANDPA

He's probably got a lot of letters to read, Annabel.

ANNABEL

Fair enough, but that's his job. I mean, Dad, you've always given money to local environmental causes and you're a Conservative.

GRANDPA

And you're a playwright who wants to bring our attention to this. But how much of your own money did you contribute to protect Brome Lake this year? (*pause*) Did you pay your dues to Renaissance Lac Brome?

ANNABEL avoids his gaze.

GRANDPA

Because I give them money every year to look after this lake that we all enjoy, but I can tell you that most people around here don't. They would rather spend their money on a speedboat or taking their family out to dinner. So this is my concern – that we talk and talk and talk and ask the government to do all sorts of things – but the government is in debt, Annabel. So who's going to pay the bills for all these things? More often than not it's the capitalist pigs like me. (*pause*) I love you, but I'm going to bed. (*taking an envelope off the Christmas tree*) There's one last envelope in the tree.

He gives it to ANNABEL.

GRANDPA

Merry Christmas, darling. Spend it wisely.

SCENE 13

Later that evening. ANNABEL is on her laptop.

ANNABEL

Dear Diane,
I have no idea where you are. Hoping I might find you on Facebook. Please get in touch if you get this message.

Sound of notification alert on Facebook Messenger.

Facebook Messenger conversation between CAROLINE and ANNABEL.

CAROLINE

Online on Christmas Eve?

ANNABEL

Guilty. What's your excuse?

CAROLINE

You're asking the Jew? Just back from the movies.

ANNABEL

Anything good?

CAROLINE

This Is 40. I wish. Started looking into the IISD.

BEATRICE enters and watches her mother.

ANNABEL

And?

CAROLINE

A bit opaque, which doesn't mean they're hiding anything. It's just an NGO with many offshoots. Kinda hard to track revenue.

BEATRICE

What are you doing, Mama?

ANNABEL

Oh my goodness, you startled me. What are you doing out of bed?

BEATRICE

I'm thirsty.

ANNABEL

(*to CAROLINE*) Gotta go. Can you keep digging? Especially for any connection to oil?

CAROLINE

Dog on a bone. Merry Christmas!

ANNABEL

XO. (*to BEATRICE*) Let me get you some water.

While BEATRICE is guzzling a very tall glass of water ALEX enters and turns on the TV to check the weather.

JANICE DEAN and Fox NEWS ANCHOR enter.

JANICE DEAN

The wicked winter storm Euclid swept east across North America Wednesday, creating a post-holiday travel *nightmare* from California to New Hampshire and up to Toronto and Montreal.

The NEWS ANCHOR shakes his head at the bad weather.

JANICE DEAN

The whiteout came a day after a Christmas storm unleashed heavy snow, deadly winds, and even some tornadoes on the U.S. Gulf Coast, killing at least three people.

NEWS ANCHOR

Holy shit, J.D.

JANICE DEAN

Nearly two thousand flights have been cancelled and ten thousand delayed, many at Dallas–Fort Worth, Philadelphia International, and Montreal's Trudeau International.

ANNABEL

(*to ALEX*) I know the timing's not great. But are you okay if I spend some of this (*referring to envelope from her GRANDPA*) to fly to Edmonton?

SCENE 14

DR. DAVID SCHINDLER speaks with ANNABEL in his office at the University of Alberta in Edmonton.

ANNABEL

I'm just wondering why Diane left so suddenly.

SCHINDLER

Well, she had to get back to her Ph.D. studies or she would have put herself in jeopardy with the university.

ANNABEL

Is that the only reason?

SCHINDLER

As far as I know.

ANNABEL

Have you heard from her since she left?

SCHINDLER

I'm sorry but I don't feel at liberty to speak about that.

ANNABEL

She told me you are her adviser and mentor.

SCHINDLER

Yes. (*pause*) If you're looking for someone to speak with about the ELA, though, there are other scientists still doing work at the site over the winter.

ANNABEL

Anyone in particular I should speak with?

SCHINDLER

I'd get in touch with the group from Trent University doing the silver nanoparticle experiment. They'd have some information for you.

ANNABEL

Thank you.

SCHINDLER

You're welcome.

ANNABEL

And in terms of the IISD, have you spoken with Hank Venema since the event in Winnipeg? I've phoned him several times but he hasn't returned my calls.

SCHINDLER

Well, he may not be free to speak about the negotiations.

ANNABEL

Why not?

SCHINDLER

Our government seems to want to have everything carried on behind closed doors. Last time I've seen anything like this was back in the early 1970s when scientists from the former Soviet Union used to have to get special permission to go to meetings and there was always a KGB agent with them.

ANNABEL

But we're not talking about enriching uranium here. We're talking about freshwater science.

SCHINDLER

Well, they're afraid of any science that might throw a monkey wrench in their agenda. But by muzzling science they're actually increasing suspicion about their agenda. What generates fear is not the known but the unknown.

ANNABEL

Which is probably why so many people fear the Harper government.

SCHINDLER

I would say so.

STEPHEN HARPER enters.

ANNABEL

Do you fear them?

SCHINDLER

Well, I'm going to retire later this year. Nothing much to lose. Plus, I've had the one-on-one meetings with this government. The Minister of the Environment Peter Kent – he certainly knows my feelings about all this very well.

ANNABEL

Why is that?

SCHINDLER

Because his department pretty well had to drop a lot of other things to be able to pick up their monitoring of the Athabasca River after my studies out there.

ANNABEL

Which studies?

SCHINDLER

Diane didn't tell you about this?

ANNABEL

No.

SCHINDLER

Well, you know, for years the Alberta government had been solely responsible for monitoring the environmental impact of the oil sands.

ANNABEL

Okay.

SCHINDLER

And they were putting out the statement that, according to their data, the oil sands industry is having no effect on the Athabasca watershed. And, uh, that just didn't make much sense to me. And I did a study in 2008, which was published in 2010, that showed, indeed, industry was polluting the river significantly.

ANNABEL

Oh my God.

SCHINDLER

But then the Alberta government denied my challenge to their results. They said, "Oh, our studies are correct." So I talked Peter Kent into getting some panels of experts to look at my data. Five different panels were appointed, and they all said my data was correct.

ANNABEL

So this was in 2010?

SCHINDLER

Late 2010, that's right.

ANNABEL

About a year before the federal government announced the defunding of ELA?

SCHINDLER

About that, yes.

ANNABEL

Are you trying to tell me that ELA may have been cut in retaliation for your study?

SCHINDLER

Uh, I have seen no evidence that would lead me to that conclusion ... yet.

SCENE 15

ANNABEL watches a video of STEPHEN HARPER on her laptop on the plane back to Montreal.

Sound of applause.

Prime Minister STEPHEN HARPER makes an address to the Canada–U.K. Chamber of Commerce. London, July 14, 2006.

STEPHEN HARPER

Ladies and gentlemen, one of the primary targets for British investors today has been our booming energy sector. They have recognized Canada's emergence as a global energy powerhouse – the emerging "energy superpower" our government intends to build.

It's no exaggeration. We are currently the fifth-largest energy producer in the world. We rank third and seventh in global gas and oil production, respectively. But that's just the beginning. An ocean of oil-soaked sand lies under the muskeg of northern Alberta – my home province.

The oil sands are the second-largest oil deposit in the world, bigger than Iraq, Iran, or Russia; exceeded only by Saudi Arabia. Digging the bitumen out of the ground, squeezing out the oil, and converting it into synthetic crude is a monumental challenge. It requires vast amounts of capital, Brobdingnagian technology, and an array of skilled workers.

In short, developing Canada's oil sands is an enterprise of epic proportions, akin to the building of the pyramids or China's Great Wall, only bigger.

SCENE 16

Back at home, ANNABEL's smartphone starts to ring.

BEATRICE

Mummy, we finished our research assignment on Stephen Harper.

ELLA

Did you know he really loves cats?

ANNABEL

Girls, this is my scheduled call.

ELOISE MITCHELL, Department of Fisheries and Oceans, enters.

ANNABEL

Hello?

ELOISE MITCHELL

Hello, Annabel. This is Eloise Mitchell phoning with communications at Fisheries and Oceans Canada.

ANNABEL

Yes. Hello, Eloise.

ELOISE MITCHELL

I'm here to connect you for your twenty-minute call with Dave Gillis, assistant deputy minister, Fisheries and Oceans Canada. Are you ready?

ANNABEL

Oh yes, I'm ready.

ELOISE MITCHELL

Before I pass you over to him, I just want to remind you, as we agreed to previously, there were some political questions you sent us that we couldn't necessarily offer you. But there is a wide spectrum of questions we can take.

ANNABEL

I understand. I'll try to stick to the script.

Call is transferred.

DAVE GILLIS

How are you today?

ANNABEL

I am very well, thank you. How are you, Mr. Gillis?

DAVE GILLIS

I'm fine, thank you.

Pause.

ANNABEL

Well, I guess I'll jump right in since we don't have much time. Mr. Gillis, when did you first become aware that public funding to the ELA would be cut?

DAVE GILLIS

It was announced in the budget of 2012.

ANNABEL

So you knew before that then?

DAVE GILLIS

Well ... before that ... I mean, really, you know, very, very, very, *very* few people actually know what is in the budget before it is read in the House.

ANNABEL

So you didn't know?

DAVE GILLIS

No, not the final decision.

ANNABEL

So if you – the assistant deputy minister of fisheries and oceans – didn't know, who was involved in making that final decision?

DAVE GILLIS

Uh, I have to be a little careful here about the process. At some point here it goes behind, you know, the veil.

ANNABEL

If you could take me up to the veil then.

DAVE GILLIS

Uh, well, in some generalities I can talk about that. Uh, there, of course, is this exercise which everyone knows about, called the Strategic and Operating Review that was conducted by all departments and agencies of the federal government in 2012.

ANNABEL

When you were all asked to cut between 5 and 10 percent of your budgets?

DAVE GILLIS

Yes, it wasn't just Fisheries and Oceans.

ANNABEL

I know, I distinctly remember it happening at Heritage Canada too.

DAVE GILLIS

Right. So you know, we did that. We were a small team of, uh, public servants who were vested with looking for those, uh, options.

ANNABEL

And so the ELA would have been one of the options you submitted for cuts?

DAVE GILLIS

That someone, uh, submitted, yes.

ANNABEL

But I assume some choices had to be made based on ... not just how funds, when cut, would contribute to deficit reduction – but also based on the value that each program contributes to society. (*pause*) In other words, why would it be worth cutting quite a small budget – two million dollars – for something that delivers such an invaluable service not just to Canada but to the world? And here I'm referring to the billions of dollars that have been saved over the past forty years because of the research that ELA provided about blue-green algae in the 1970s and about sulphur-dioxide emissions from U.S. smokestacks that were killing our lakes in the 1980s.

Pause.

DAVE GILLIS

Uh, I'm sorry, Ms. Soutar. It's actually quite difficult for me to say anything further.

SCENE 17

ANNABEL interviews DR. MAGGIE XENOPOULOS in her office at Trent University in Peterborough, Ontario, where MAGGIE is a professor of aquatic science.

MAGGIE

Yeah, nobody wants to go on the record from their side.

ANNABEL

Why not?

MAGGIE

I don't know. Everyone is being censored.

ANNABEL

But censored by whom? I'm the taxpayer. I'm the client here. Public officials should have to defend this decision to me.

MAGGIE

Well, I guess there's not enough of us taking them to task.

Pause.

ANNABEL

Maggie, you were working on a whole-lake experiment when the funding cut to the ELA was announced. David Schindler said on sliver nanoparticles?

MAGGIE

Yeah, it's an emergent contaminant. Silver nanoparticles might be in your socks, your underwear, so when you wash them, this stuff winds up directly in our waterways. We have no idea what it does to our ecosystem. That's what we wanted to do with our study. We got almost a million dollars from NSERC, actually, to do this, which we can't use now because we don't know the future of ELA.

ANNABEL

But if you have your own research money, can't you ...?

MAGGIE

The grant was given to do a whole-lake experiment. Without ELA, the whole thing falls apart.

ANNABEL

Maggie, how do you explain the ELA funding cut?

MAGGIE

There is no reason that I understand. It makes no sense to say it saves money.

ANNABEL

When I met with David Schindler last week, he told me about his oil sands studies in 2010.

MAGGIE

So you've heard the conspiracy theory.

ANNABEL

I have. But frankly I'm reluctant to accept it.

MAGGIE

Why?

ANNABEL

It just seems too simple.

MAGGIE

I don't usually buy them either.

ANNABEL

But?

MAGGIE

But I do think Dave Schindler had something to do with it.

ANNABEL

Why?

MAGGIE

Because I've known him for a long time. He was my adviser too, and he's ... He can be a shit disturber.

DR. PAUL FROST suddenly pops his head into MAGGIE's office.

PAUL

Hi.

MAGGIE

Hi.

PAUL

Can we talk?

MAGGIE

Yeah.

Pause.

PAUL expects MAGGIE to come out into the hall but she doesn't budge.

MAGGIE

What?

PAUL

I just got the call. The weather is clear so ... it's a go for Sunday.

MAGGIE

Great! This is Paul.

ANNABEL

Hi. Annabel.

PAUL

Hi.

MAGGIE

We're working together on the silver nanoparticle project, and he also happens to be my husband.

ANNABEL

Oh!

MAGGIE

And one of Canada's top limnologists. He's heading up to the ELA on Sunday to take some winter samples. Maybe you two should talk. (*starting to leave*) I have a meeting.

PAUL

(*reluctant*) Sure.

MAGGIE

Annabel's a playwright. She's interested in the ELA. And she needs to see it before it closes. (*to ANNABEL*) Right?

ANNABEL

Uh, right.

MAGGIE

She should go with you.

MAGGIE gives PAUL a peck on the cheek and leaves.

PAUL

Uhhh ...

ANNABEL

Uhhh ...

SCENE 18

The Experimental Lakes Area research station in North Western Ontario. ANNABEL and PAUL stand in the middle of a frozen lake. PAUL carries a bag with water sampling equipment.

ANNABEL

This is the most beautiful laboratory I've ever seen.

PAUL

Yup.

ANNABEL

So quiet.

PAUL

Not for long, I'm afraid.

PAUL pulls the starter cord on an imaginary ice drill.

Sound of the ice-drill motor blasting to life.

PAUL finishes drilling the hole.

PAUL

See how the pressure builds up under the ice?

ANNABEL

Yeah, the water just spurts up – like oil.

PAUL

So I'm just gonna ask you to hold this.

PAUL hands ANNABEL a sample container with a funnel on top and then lowers another bottle on the end of a fishing line into the hole in the ice.

PAUL

We're just going to take water with this bottle, and then we're gonna pour it in there.

ANNABEL

You're not going to pour freezing water all over my hand, are you?

PAUL

Yeah. You'll get used to it. Aw shit! We just hit the bottom.

ANNABEL

Already?

PAUL

Yeah, it's only, like, a metre here.

ANNABEL

But we're so far from shore.

PAUL

Yeah, you never know with a lake. Oh God, now there's crap all the way up the line. We'll have to let the sediment die down.

ANNABEL

Can't you just drill another hole?

PAUL

No. We'll just wait.

He fills up one of his empty sample bottles with freezing lake water and offers the bottle to ANNABEL.

PAUL

Here – taste this.

ANNABEL drinks.

ANNABEL

So good. I can't remember the last time I drank from a lake.

Lighting change to suggest time passing.

Later that day, ANNABEL and PAUL stand outside, looking out over a lake at sunset.

PAUL

This is Lake 240.

ANNABEL

Incredible.

PAUL

My kids would never believe what it looks like now. They were here last summer with me, jumping off this dock into the water. My youngest son loves bugs so he was just grabbing the leeches out of the water. (*laughing*) Agh! Leeches!

ANNABEL

(*laughing*) I have a daughter who loves slugs.

PAUL

Yeah.

ANNABEL

I spent my summers growing up by a lake in Quebec.

PAUL

Oh really? Which one?

ANNABEL

Brome Lake.

PAUL

I know Brome Lake.

ANNABEL

You do?

PAUL

Yeah, with the island in the middle of it.

ANNABEL

Yeah! Eagle Island. My family has been spending summers there for four generations. My paternal grandfather bought the property in 1941 as a summer retreat from their home in Asbestos. My grandfather worked for the mining company.

Lighting change to suggest time passing.

PAUL

This is where my family spent the summer last year when we started our silver nanoparticle experiment. It's one of the cabins that's reserved for scientists with kids. One of the few places Maggie and I could work and be with our kids at the same time.

ANNABEL

That sounds amazing.

PAUL

That's the volleyball court over there. It used to be, like, every Wednesday night was the big volleyball tournament. You know, scientists can be pretty competitive. We'd have special events every week, like a wine and cheese and everybody gets dressed up, everybody gets pretty hammered on wine. We have this variety night – that's actually a stage over there under that snowbank – people do skits and ... and play music and ...

PAUL gets a bit choked up.

Lighting change to suggest time passing.

Later that evening, ANNABEL and PAUL stand in the chemistry lab looking at huge data charts on the wall.

PAUL

These are the charts that show how long they've been collecting data. So from the 1960s on they were looking at the question of when the ice would come off each spring. So a decade ago, the ice was melting in the middle of May. This year? They're telling us it's going to come off at the end of March. It's, like, in ten years to have such a drastic change? How is that possible? But it's only because we have a long-term record that this starts to have, like, any sort of meaning. It's *continuous* records since the 1960s. This kind of data is so rare.

ANNABEL

So when we're talking about climate change data, this is the shit.

PAUL

Fifty years of continuous data on climate change and they want to stop now?

ANNABEL

Shows the disconnect.

PAUL

Exactly.

SCENE 19

House of Commons in Ottawa. Members of Parliament debate an Opposition Day Motion to reverse federal funding cuts to the ELA. March 20, 2013.

KENNEDY STEWART (*NDP MP for Burnaby-Douglas*)

Let it be moved that, in the opinion of this house, the federal government should maintain support for its basic scientific capacity across Canada, including immediately extending funding to the world-renowned Experimental Lakes Area to pursue its unique research program!

Different party members of the House of Commons shout "Hear, hears" or heckles according to their affiliations. This continues throughout the scene.

KENNEDY STEWART

Mr. Speaker, science is not test tubes or data sets or microscopes or space stations, but a method by which we explore and attempt to explain our world. Without a strict adherence to the scientific method, we do not generate science but mere propaganda.

DAVID MCGUINTY (*LIB MP for Ottawa South*)

Mr. Speaker, are we to understand that the government cannot afford the two million dollars per year to fund freshwater research being done at the ELA? If so, Canadians should know that since the Conservative Party arrived in power, it has spent six hundred million dollars on advertising.

MICHELLE REMPEL (*Parliamentary Secretary to the Minister of the Environment, CPC MP for Calgary Nose Hill*)

Mr. Speaker, every single one of these facts that my colleagues have stated have been out of context, misquoted. They are patently fear-mongering.

LAURIN LIU (*NDP MP for Rivière-des-Mille-Îles*)

Monsieur le président, les politiques publiques doivent s'appuyer sur la science.

Simultaneous voiceover translation.

ENGLISH TRANSLATION (*voiceover*)

Mr. Speaker, political discourse should be based on science.

LAURIN LIU

Plutôt que sur des préjugés idéologiques –

ENGLISH TRANSLATION (*voiceover*)

Instead of on ideological bias –

LAURIN LIU

En instaurant un régime de terreur –

ENGLISH TRANSLATION (*voiceover*)

In establishing a regime of terror –

LAURIN LIU

Le gouvernement conservateur tente de faire taire les scientifiques qui pourraient le contredire.

ENGLISH TRANSLATION (*voiceover*)

The Conservative government is attempting to silence scientists who might contradict their agenda.

LAURIN LIU

C'est inacceptable.

ENGLISH TRANSLATION (*voiceover*)

This is unacceptable.

GARY GOODYEAR (*CPC MP for Cambridge*)

Mr. Speaker, we are very strong in this country in basic research and we intend to stay there. Where we could do a little better is on the *commercialization* end of that knowledge. We have an obligation to do that. If we are serious about saving the environment, and if we are serious about improving quality of life and saving lives, we must move those discoveries out of the laboratories, build those products in our factories, and sell them to the –

ANONYMOUS LIBERAL MP

Boo! Booooo! Boooooooo!

Pause.

GARY GOODYEAR

And sell those products to the living rooms and hospitals of the world. As the prime minister has often said, science powers commerce.

ACTING SPEAKER OF THE HOUSE (*Barry Devolin, CPC MP for Haliburton-Kawartha Lakes-Brock*)

It being 6:15 p.m., and today being the last allotted day for the supply period ending March 26, 2013, it is my duty to interrupt the proceedings. All those in favour of the motion will please say yea.

LIB/NDP/IND

Yea!

ACTING SPEAKER OF THE HOUSE (*Barry Devolin*)

All those opposed will please say nay.

CPC

Nay!

ACTING SPEAKER OF THE HOUSE (*Barry Devolin*)

In my opinion the nays have it. I declare the motion defeated.

SCENE 20

At home. ANNABEL, ELLA, BEATRICE, and ALEX switch off the Cable Public Affairs Channel (CPAC) where they have just watched the House of Commons debate and ELA motion being defeated.

BEATRICE

The popcorn was really good, Mama. Thanks a lot.

ALEX

Bring the bowl to the sink and wash it, Bibi.

ELLA

Will the ELA get another vote, Mummy?

ANNABEL

No.

ELLA

But I thought you talked to our mayor and he said –

ALEX

Our federal MP, Ella.

ELLA

– our federal MP, and he said he was going to vote for the ELA.

ALEX

Marc Garneau is only one person, Ella.

ELLA

But he's an astronaut!

ANNABEL

(*looking at her smartphone*) Oh my God.

ELLA

What?

ANNABEL

Diane is back in action.

ALEX

How do you know?

ANNABEL

She's posting on Twitter about the vote. This is her first tweet in months.

DIANE enters and speaks onstage.

DIANE

(*Twitter post*) I can't even find the words to describe how crushing this outcome is. For ten months, we tried every democratic tool available. My last hope. We lost.

ANNABEL starts typing on her smartphone.

BEATRICE

What are you doing, Mama?

ANNABEL emails DIANE.

ANNABEL

(*email*) Diane, Can we talk? Where are you?

DIANE

(*email*) In Winnipeg. But I'm heading up to the ELA tomorrow to make some noise.

ANNABEL

(*to ALEX*) Diane's going up to the ELA!

DIANE

But keep that to yourself, please.

ANNABEL

(*email*) My lips are sealed. Can I ... come with you?

Pause.

DIANE

(*email*) Call me on my cellphone right now.

ANNABEL calls DIANE and puts her on speaker phone. ALEX rushes to turn on the recorder and then listens in.

DIANE

I should warn you that I've arranged to have a film crew come to ELA.

ANNABEL

Okay.

DIANE

They're from CBC TV.

BEATRICE

TV!

ALEX

Shh!

ANNABEL

CBC out of Winnipeg?

DIANE

Yeah but it's national.

ANNABEL

Wow. That's cool.

DIANE

We have to get some footage of ELA before it closes next week.

ANNABEL

Okay well, we're totally prepared for whatever.

Pause.

In another area of the stage, a WORKMAN enters and starts dismantling a structure.

ANNABEL

Diane? Diane? Are you still there?

DIANE

(*choked up*) They've already started gutting the cabins, Annabel.

ANNABEL

What? The cabins?

DIANE

The family cabins. There was still peoples' personal stuff in there –

ANNABEL

How do you know about this?

DIANE

I can't tell you. Can you get up to the ELA site by Thursday morning?

ANNABEL

Uh ...?

Sound of an airplane.

SCENE 21

CAMERON (a reporter), ANGELA (a producer), and JOHN (a cameraman) have arrived to shoot an interview with DIANE for CBC TV's The National. *ANNABEL and ALEX observe. March 22, 2013.*

JOHN

Okay, I'm ready when you are.

CAMERON

Okay, so –

DIANE

Can I make a statement? Before we do this?

CAMERON

Yeah sure. We'll just do the interview. And then you can –

JOHN

I just need you to count to ten for your mic now, please.

DIANE

Uh, one, two, three, four, five –

While DIANE keeps counting, ANNABEL speaks aside to ALEX.

ANNABEL

Diane told me she was thinking of occupying ELA.

ALEX

Occupying?

ANNABEL

Yeah, apparently she's already stockpiled barrels of food and water in one of the cabins.

JOHN

Okay. Good. Okay, look towards Cam. Yeah, that's great.

CAMERON

Talk to me.

JOHN

Okay. I'm rolling, guys.

CAMERON

All right. So, Diane. ELA. Um, what's the ... *situation*?

DIANE takes a deep breath.

Cut to later in interview.

JOHN

Okay, we're rolling.

CAMERON

So ... to the *lay person*, what does it mean if scientists can't come here to do their work anymore? What are we losing *actually*?

DIANE

We're losing –

Cut to later in interview.

JOHN

Rolling.

ANGELA

Now I don't know how many summers you've spent out here, but probably a few?

DIANE

Ten years.

ANGELA

Ten years of *your life*. When you think about it, a week from now, not being allowed to come here anymore, how does that make you feel?

DIANE

It just breaks my heart ... It's just ... devastating. This place is like no other in the world. We need this place so badly. Um, there's a lot of things we can do without, but water's not one of them.

JOHN

Beautiful.

ANGELA

I think that's everything I wanted to ask. Cam?

CAMERON

That's everything I got.

ANGELA

Sorry, what lake is this again, Diane?

DIANE

This is Lake 239.

ANGELA

Two-three-nine.

CAMERON looks at his smartphone.

CAMERON

(*chuckling*) Oh my God. Check out this story we missed in Winnipeg today.

ANGELA

What?

CAMERON

A baboon is running loose at the MTS Centre.

ANGELA

Get out.

DIANE

Um, could I? I just have one statement that I'd like to make.

Pause.

CAMERON

Yup.

DIANE

For the camera, just one thing I'd like to say directly if that's okay?

ANGELA

Sure. John?

JOHN

Uh ... well, okay. Yeah, this is a good enough spot right here. Can I just get you to count again?

DIANE

One, two, three, four –

JOHN

Okay, you're good to go.

DIANE

In two days the Experimental Lakes –

JOHN

Okay, but just keep looking at Cam on this side, not directly at the camera.

DIANE turns to the live theatre audience and speaks to them directly.

DIANE

(*to audience*) In two days the Experimental Lakes Area will no longer be a public-science program under the purview of the Government of Canada. What does this mean to you, the Canadian taxpayer? It means that the ELA's research priorities will cease to reflect *your* priorities or *your* interests. On March 31, despite the fact that three out of four Canadians have asked the federal government *not* to cut its funding, the ELA will either die – or be transferred to an NGO whose private sector donors may represent the very fucking industries whose environmental impact on our water ELA is supposed to be monitoring! (*pause*) No response.

SCENE 22

Later, ALEX, DIANE, and ANNABEL debrief after news crew have left.

DIANE

Did I screw this whole thing up by overpoliticizing it? As a citizen of this country, I didn't agree with a decision that was made by my government. So I asserted my free democratic right to oppose them on that decision. Is that perceived as *radical* behaviour? And if it is, then in my opinion *that* perception is at the root of all this – of why it's all so rotten.

ALEX

You had to keep the story alive for the public. No one was doing that as effectively as you.

ANNABEL

Yeah but if the only way to be effective in terms of getting attention also hardens your opposition against you, then ...

DIANE

Exactly. A lot of people think I fucked up royally.

ALEX

No.

ANNABEL

You didn't fuck up.

DIANE

Some of my ELA colleagues who are working on the transfer to IISD – they told me I should keep quiet or the government would kill the deal. To have close, personal colleagues say you're doing more harm than good, after everything I've done?

ANNABEL

Diane –

DIANE

Maybe a better strategy would have been to have some private conversations with Conservative MPs. But you know what? I didn't want to do that because it's an important story to tell Canadians.

ANNABEL

It is.

ADÈLE HURLEY enters.

DIANE

Which is what Adèle said to me from the beginning.

ANNABEL

I'm sorry, Adèle?

DIANE

Adèle Hurley. She's the director of the water program at the Munk School for Global Affairs.

ANNABEL doesn't know it.

DIANE

At U of T?

ANNABEL

Okay.

DIANE

Like, she coined the term "acid rain."

ANNABEL

Wow.

DIANE

Early on, when I was developing my media strategy, I travelled to Toronto to meet with her. I'm sitting there with a director at the Munk School for Global Affairs, and she just looks at me. And she's, like –

ADÈLE HURLEY

This country is being run by a tyrant.

DIANE

And I'm, like – I'm shocked, you know?

ADÈLE HURLEY

You wanna win this fight for ELA? You can't play nice.

DIANE

And ... I'm a nice, quiet person. I'm usually nice.

ADÈLE HURLEY

You have to make this: "Harper killed ELA. Harper: Bad."

DIANE

Like, that kind of language.

ADÈLE HURLEY

During the acid-rain days, all the scientists were talking about "acidic deposition" and how that "might translate into changes in fish populations." Well, I'm calling it *acid rain.* I know that's wrong. But I'm calling it acid rain. And when I go to those fishermen: "That acid is, like, *killing* your bass."

DIANE

And I'm just: "Every rule I know as a scientist is being thrown out. I was trained to be very objective, and all of a sudden, you're telling me ...?" But I was, like, if anyone knows, she knows.

ANNABEL

Because she actually got something important done.

DIANE

Exactly. So I took her advice. I attacked Harper. I attacked Kent. I attacked Ashfield. All of them. (*choked up*) Did I jeopardize ELA by doing that? Maybe I did. I don't know.

ANNABEL

You had to get the story out. Diane, I wouldn't be here if I hadn't met you. None of us would.

DIANE

Someone once used the analogy: ELA is but one tree in the forest. Some people are trying to save the one tree, whereas I think I was using that one tree to try to save the whole forest.

MAUDE BARLOW enters and reads an announcement from the Council of Canadians website.

MAUDE BARLOW

March 31, 2013
Federal funding for the Experimental Lakes Area ends today. Tonight the Council of Canadians held a candlelight vigil at the ELA in northern Ontario to mark this dark moment in our country's environmental legacy.

SCENE 23

Easter 2013. GRANNY and GRANDPA's house at the Lake Nona Golf and Country Club in Florida. GRANNY, GRANDPA, ANNABEL, ELLA, and BEATRICE paint hard-boiled eggs out on the patio in the bright sunshine. ALEX hits golf balls on the practice range nearby.

ANNABEL

I have to speak with someone in the Conservative Party about this, Dad.

BEATRICE

Do you like my egg, Mama?

ANNABEL

It's beautiful, darling.

GRANDPA

I thought all that was over now. Didn't the deadline already pass?

ANNABEL

Yes – for keeping ELA a federally funded research site. But the government might still negotiate a transfer to an NGO. And it's all happening behind closed doors – no one knows for sure. Do you think some of the people you know in the party might talk with me about it?

GRANDPA

Well, I don't know. It seems there's a lot of bad blood over this ELA thing.

ANNABEL

There is bad blood. But it's being totally misplaced. The ELA scientists I've spoken to, they're not against industry.

GRANDPA

Well, I think people in industry recognize that, darling. And I think they recognize that they have to listen to environmental scientists. I mean, when you develop a mine today, you can't even get a permit to develop that mine unless you agree to restore it to the original state.

ANNABEL

Which is a good thing.

GRANDPA

And you've got to protect against any kind of leakage of any kind of toxic materials. I mean, when you're talking about developing a new mine today, you're talking about a minimum of a ten-year commitment before you start to get any money back.

GRANNY

Well, Ian, mining puts a huge strain on the environment –

GRANDPA

But all I'm saying, Helgi, is that industry *is* playing by the rules –

GRANNY

– even if they do clean it up.

GRANDPA

They have all of these ... protective tailing pools –

GRANNY

And then also the processing –

GRANDPA

And –

GRANNY

They use cyanide solutions and other toxic chemicals.

GRANDPA

I *know*.

GRANNY

I mean, they still use cyanide to –

GRANDPA

But, *Helgi*. You don't get a permit anymore to mine –

GRANNY

No, that I understand –

GRANDPA

Unless you put in place all of these, frankly, very costly protective measures – measures that I think our prime minister is trying to loosen up a bit so we can get some business done in this country.

Pause.

STEPHEN HARPER enters. He stands in the background or on side, not in a spotlight.

ALEX enters from the golf range.

ANNABEL

Dad, I'm not trying to write a story that's, like, Harper the Big Bad Monster here.

GRANDPA

Well, everybody already believes that anyway.

GRANNY

Well, maybe he needs to learn to be a bit more communicative.

Pause.

GRANDPA

He's not a bad person, Helgi. I mean –

GRANNY

I'm not *saying* –

GRANDPA

He's a human being.

ALEX

(*to ANNABEL*) "Cognitive dissonance."

GRANNY

I'm not saying he's a bad person –

GRANDPA

And he'll make all kinds of mistakes.

ALEX

Apparently this is what Naomi Klein's next book is all about.

GRANDPA

And he certainly has an agenda –

GRANNY

Please don't say the name Naomi Klein in this house.

GRANDPA

But that agenda is to make sure that economic growth is alive and well –

GRANNY

Dreadful woman.

GRANDPA

I mean, we should be *thanking* Stephen Harper. Because that's the source of all of the spending we love doing in Canada. I mean, we basically balance our trade deficit by selling the world raw materials, and a huge part of that is oil. It's the centrepiece of Canadian policy, Annabel. It's *gigantic* in terms of wealth creation for all Canadians.

GRANDPA coughs violently. ELLA brings him a glass of water.

Pause.

GRANDPA

Anyway, I'll call Max Bernier when we're back in Montreal and see if he'll do an interview with you.

ANNABEL

I really appreciate that, Dad.

GRANNY

You should call John, Ian.

GRANDPA

Yeah, I'll call John. He hits me up for a donation every year. That's got to be worth something.

GRANDPA shows BEATRICE his painted egg.

GRANDPA

Whaddya think, Bibi? Does that look like Humpty Dumpty?

BEATRICE

Where's his crack?

GRANDPA ponders his egg.

ELLA

I think it's beautiful, Grandpa. (*pause*) Are you going to be out of the hospital for my birthday, Grandpa?

GRANDPA

That's the plan, darling. That's the plan.

STEPHEN HARPER hands GRANDPA a document – the speech that JOE OLIVER will deliver in the next scene.

SCENE 24

JOE OLIVER, federal minister of natural resources, makes a keynote address to the University of Calgary Extractive Resource Governance Policy Program. Calgary, Alberta, April 18, 2013.

DIANE listens.

Meanwhile, ANNABEL and ALEX are at the Orlando airport returning to Montreal.

ANNABEL

Where are the girls?! We're going to miss our flight!

Sound of an airplane taking off.

JOE OLIVER

Ladies and gentlemen, my message to you today is that Canada stands at a pivotal moment –

As soon as she arrives back home, ANNABEL is anxious to leave again.

ANNABEL

Alex, I've got to go back to Winnipeg to speak with someone involved with the transfer.

ALEX

Go!

JOE OLIVER

A time when the actions we take as a country will either set the course for future growth or consign Canadians to watching the opportunities pass to others. The stakes are enormous.

ANNABEL meets with DR. MIKE PATERSON, a senior ELA scientist in the Department of Fisheries and Oceans, in Winnipeg.

MIKE

(*to ANNABEL*) Um, are you recording this?

JOE OLIVER

I do not need to tell any of you how important natural resources are to our economy. You all know the numbers today.

MIKE

I ... I just can't go on the record. I still work for DFO.

ANNABEL

I promise that whatever I record won't see the light of day until 2015.

JOE OLIVER

But what about for the future? Over the coming decade, as many as six hundred major resource projects are planned worth more than 650 billion dollars and representing more than a million new jobs for Canadians.

MIKE

The federal government is still holding the reins on the ELA transfer, Annabel. And they've made it clear: They're going to be watching us closely and deciding the timing of everything.

JOE OLIVER

This is nation-building, and it is nation-building on a grand scale, comparable to the building of the great railroads in the nineteenth century.

ANNABEL

But don't they just want ELA off their balance sheet? I mean, why would they *delay* a transfer after all this?

MIKE

Because they are extremely angry about how much heat they've taken about this in the media.

DIANE

(*interrupting JOE OLIVER's speech*) But, what kind of nation are we building, Mr. Oliver?

ANNABEL

(*to MIKE*) You're referring to what Diane's been doing?

JOE OLIVER

At this pivotal moment, however, we are hearing loud objections from those who oppose virtually every form of resource development –

DIANE

Why did your government axe the National Roundtable on the Environment and the Economy if not because you perceive environmental –

JOE OLIVER

– whose opinions are not based on facts.

DIANE

Facts?

ANNABEL

Mike, is the IISD going to be able to preserve the integrity of ELA's science if this transfer happens?

MIKE

Yes.

ANNABEL

How do you know that for sure?

MIKE

Because I have pushed Hank Venema at the IISD hard on exactly that issue. And I gotta tell you, he has passed every smell test with me.

JOE OLIVER

My friends, Canada was not built by naysayers.

ANNABEL

So what do you need now, Mike? I want to help you.

DIANE

(*to JOE OLIVER*) You're the ones who say no to any research that doesn't buttress your ambition to turn this country into the battery pack for the world –

MIKE

We need money, a lot of it, as soon as possible.

JOE OLIVER

It was built by women and men who dared greatly.

CLAUDIA BARILÀ enters. ANNABEL answers a call on her smartphone.

ANNABEL

(*into phone*) Claudia!

CLAUDIA

What's up, girl? Your message sounded so serious.

JOE OLIVER

Who responded to the challenge of nation-building with the eloquence of *action*.

ANNABEL

(*rushed*) Claudia, I hope you don't take this as opportunistic but I need you to talk to Guy, or someone else at the One Drop Foundation right now.

DIANE

(*to JOE OLIVER*) Why haven't you taken the *action* you promised Canadians back in 2006 to promulgate regulations on carbon emissions?

CLAUDIA

(*laughs*) Right *now*?

CLAUDIA exits. JOE OLIVER and DIANE text starts to overlap.

JOE OLIVER

Just imagine how different our history would have been if we had said no to building the great railroad across an unforgiving terrain.

DIANE

When a government goes to such extreme lengths to promote such a short-sighted and narrow interest as a single-polluting industry, it's an affront to the democratic –

JOE OLIVER

Through swamps and streams, hard granite and high mountains!

DIANE

You are not managing for the long-term view.

JOE OLIVER

An undertaking that spanned a continent, amazed the world, and made Confederation possible!

DIANE

Don't you want to make sure you get it right?

JOE OLIVER

My friends, we have the chance to turn this moment of opportunity into decades of prosperity.

DIANE

Unlike what you did with the cod fisheries?

JOE OLIVER

To answer the call of nation-building once again.

DIANE

Your environmental capital ...

JOE OLIVER

(*interrupting his speech to speak to DIANE*) WILL YOU SHUT UP?! JUST SHUT UP?! (*pause*) It's just *rude*. We're trying to build an economy here.

Actor playing JOE OLIVER transforms into GRANDPA.

GRANDPA

Annabel, we can do this. But we have to be strategic. Focused. We have to set priorities to build infrastructure so our oil and gas can reach new markets. We can develop our resources responsibly. Because overcoming challenges is what Canadians do best. It is in our DNA. It is who we *are*.

ELLA and CLAUDIA enter.

ELLA

Mummy, I need help.

CLAUDIA speaks to ANNABEL on her smartphone.

CLAUDIA

Annabel, I told Guy about it – everything you told me, but right now nothing at One Drop is exciting him at all because he has all the job cuts at Cirque to deal with.

ELLA

(*to ANNABEL*) Can you help me with my math problem?

ANNABEL

(*to ELLA*) Just give me a sec, darling.

CLAUDIA

He's like: "I'm dealing with Cirque now, Claudia! I'm dealing with my mom-and-pops company!" No ... it's ... it's a lot of stress.

ELLA

I don't understand, Mummy.

Enter ASSISTANT TO SENATOR SEIDMAN, who is calling on ANNABEL's other phone line.

ANNABEL

(*back to phone*) Claudia, can I put you on hold? Please don't go away.

ASSISTANT TO SENATOR SEIDMAN

This is Senator Seidman's office calling you back.

ANNABEL

Yes, hello. Did my father explain to you why I wanted to get in touch with Senator Seidman?

ASSISTANT TO SENATOR SEIDMAN

Yes, he did. And I left him a message this morning saying that Senator Seidman is not an expert in this field.

ELLA

I really don't understand.

ANNABEL

So I understand, but –

ASSISTANT TO SENATOR SEIDMAN

And that there might be more appropriate people you'd like to meet with.

ELLA

Mummy!

ANNABEL

Alex?! (*to ASSISTANT TO SENATOR SEIDMAN*) I'm sorry, can you just ...? (*hits Call Waiting and goes back to call with CLAUDIA*) Claudia, what about people on the One Drop board? Did you tell them the ELA needs money?

CLAUDIA

Yeah.

ANNABEL

And?

ELLA

Daddy!

CLAUDIA

It's that ... they're all pro-development, Annabel. They don't want to get behind anything that's, like, against the oil sands and all that stuff.

ELLA

Agh. Why don't I understand?

ANNABEL hits Call Waiting again and goes back to the call with ASSISTANT TO SENATOR SEIDMAN.

ANNABEL

Are you still there?

ASSISTANT TO SENATOR SEIDMAN

Not for long.

ANNABEL

(*to ASSISTANT TO SENATOR SEIDMAN*) I'm sorry. Um, the reason I asked my father to help me with this is because, you know, he *is* a Conservative Party donor.

ASSISTANT TO SENATOR SEIDMAN

Yes.

CLAUDIA calls back.

CLAUDIA

I have to get going, Annabel.

ANNABEL

Claudia, why didn't you tell them that the ELA is not an anti–oil sands organization!

CLAUDIA

I'm sorry, *ma belle*, I wish there was more I could *do*.

CLAUDIA exits.

ELLA

Mummy!

ANNABEL

Alex?! Where are you?! (*back to ASSISTANT TO SENATOR SEIDMAN*) I'm sorry, I'm not ... I'm not trying to be a bully ...I'm trying to forge a ... a new conversation.

BEATRICE enters.

ANNABEL

I realize that there has been some heated language that has been exchanged about the ELA in the past – some of it very public and well, emotional – and yet what we are talking about here is, after all, a *research* site. (*laughing desperately*) A research site to protect fresh water. A research site that right now if we could all just open up a bit –

BEATRICE

What's for supper, Mummy?

ANNABEL

– could – through federal government largesse and a bit of support from the private sector – find a new lease on life. Hello?!

ASSISTANT TO SENATOR SEIDMAN seems to have hung up.

ALEX finally enters.

ALEX

What's going on?

ANNABEL

(*gesturing to ALEX to take the kids off her hands*) You deal with this!

SCENE 25

ANNABEL sits with GRANDPA in his hospital room. He is having a chemotherapy infusion.

GRANDPA

Well, darling. You of all people should be happy when money *doesn't* talk. (*pause*) You could always go speak with someone in the oil and gas business.

ANNABEL

Who? To what end?

GRANDPA

I don't know. Someone who can convince you that pipelines are a good thing for Canada since you won't listen to your father?

ANNABEL gets up to leave. She gives her dad a kiss.

ANNABEL

I love you. I've got to get back home.

GRANDPA

I know some interesting people in Alberta.

ANNABEL

I can't afford to go to Alberta right now.

GRANDPA

Or someone right here in Montreal who works with an oil sands company? You could afford that. (*pause, then picking up his smartphone*) Anyway, I'm sending you a telephone number right now. You do what you want with it.

SCENE 26

ANNABEL talks to OIL MAN, an executive vice-president for a large oil sands company.

OIL MAN

You're lucky. You have a great father.

ANNABEL

I do. (*gesturing to her recorder*) I've already started this.

OIL MAN

Will I be identified as someone you've interviewed?

ANNABEL

Well, it's up to you. If you feel like you can be more candid with me by being anonymous, I would prefer that.

OIL MAN

Well, I can be quite candid but my company might not like it as much in a public forum.

ANNABEL

I understand.

OIL MAN

Quite frankly, the oil industry is a very conservative bunch of companies.

ANNABEL

I imagine that you guys are often unfairly vilified.

OIL MAN

Well. (*chuckling*) Yes. I mean, my personal view on the oil sands is that you can hate them all you want. I don't have an issue with that. But make sure you understand what you're hating and why you hate it.

ANNABEL

You're suggesting that people underestimate their complicity in the tar sands business?

Pause.

OIL MAN

Just one other comment. The "tar sands" – again, I don't really care what you call it – but generally people who support it call it "oil sands" and people who don't like it call it "tar sands." I don't care one way or another, but in case no one has mentioned that to you. We get a lot of people writing commentary who have never actually set foot in Alberta let alone gone to the oil sands.

ANNABEL

That's true. (*pause*) So oil has become abstract to us?

OIL MAN

Yeah.

ANNABEL

Like water.

Pause.

OIL MAN

Yeah. Actually, you're right.

Pause.

ANNABEL

Have you heard of a place called the Experimental Lakes Area?

OIL MAN

Yes.

ANNABEL

You have?

OIL MAN

I have.

ANNABEL

Do you know somebody by the name of Dr. David Schindler?

OIL MAN

Oh, I know of David quite well, yes.

DR. DAVID SCHINDLER enters.

ANNABEL

So you know about the studies he conducted around the tar ... the oil sands then?

OIL MAN

Yes, I do.

ANNABEL

How were those studies received by your industry?

OIL MAN

(*pause*) We didn't have an issue with the message. We had an issue with how the message was provided.

ANNABEL

How was it provided?

OIL MAN

David Schindler is a good scientist, but he's become an advocate.

SCHINDLER

(*to ANNABEL*) Yes, an advocate for doing what's right in the democratic process.

ANNABEL

How so?

OIL MAN

So, for example, there was a picture of him taken when he did that study holding up a picture of a fish that ... that somebody had given him in Fort Chipewyan – and he said it was a two-jawed fish – and, you know, obviously it had chemicals from the oil sands that had caused it to grow two jaws.

SCHINDLER

(*to ANNABEL*) He's got two stories garbled.

OIL MAN

Well, a couple of weeks later it came out that they'd already found it dead and it had started decaying. And that's what happens when they start to decay. There's an inner jaw and an outer jaw and they separate.

SCHINDLER

(*to ANNABEL*) The fish I held up for the press was one with a tumour.

OIL MAN

Now, David Schindler is an aquatic scientist. So I struggled that he could stand there holding this thing up and having his picture taken and put in the paper when, quite frankly, he's gotta know what he's holding up.

ANNABEL

Well, maybe he was using the media in a dishonest way in that moment –

OIL MAN

Which I think is unfortunate because he's got a lot to offer.

ANNABEL

But maybe that's the only way to get the message through these days.

OIL MAN

But the point is, he has made a decision to distance himself so far from the oil sands industry. If you want to create a dialogue about the oil sands, why would you create that ideological distance?

ANNABEL

I don't know.

OIL MAN

People ask us why we don't want to work with David Schindler. Well actually, you know, we asked. After those studies came out, my company sent him a letter saying, "Can we sit down and talk about this?" And he basically said –

SCHINDLER

No, you're all incompetent.

OIL MAN

That's not helpful, is it?

ANNABEL

No. No, it's not.

SCENE 27

ANNABEL composes a letter to Conservative Party Cabinet Ministers.

ANNABEL

Dear Prime Minister Harper,

Dear Minister Ashfield, Minister Kent, Minister Oliver,

I am writing in a spirit of open dialogue and with a deep desire to understand, from your point of view, something that just doesn't make any sense to me.

Sound of ANNABEL's smartphone ringing. Enter CHRIS ABRAHAM, calling ANNABEL from the lobby bar of a theatre in Toronto.

ANNABEL

That was fast.

CHRIS

You said you wanted my feedback.

ANNABEL

And?

CHRIS

It's a great letter. I think it carries the right balance of ideas.

ANNABEL

Okay, that's good.

CHRIS

Did, uhhh ... did you send it out to anyone yet?

ANNABEL

No.

CHRIS

Okay.

ANNABEL

If you're okay with it though, Chris, I'm going to send it tomorrow ... by email ... and by mail.

CHRIS

I'm totally fine with it. I just ... I wonder if you would be okay not including a reference to Crow's Theatre in it.

Pause.

ANNABEL

Did I ...? Where did I ...?

CHRIS

You call the play a co-production between our companies.

ANNABEL

Oh yeah. Well, it is a co-production.

CHRIS

Of course it is. I just wonder if that's really essential to mention for the purposes of your letter.

Pause.

ANNABEL

It's certainly not the main idea, no.

Lights start flashing in the theatre lobby, indicating the end of intermission.

CHRIS

Right. So given the fact that we know that ELA is such a sensitive issue. And given the fact that I'm at a pretty critical moment of our capital campaign – I mean, if it doesn't make a difference one way or another for your letter – I would just prefer that you take Crow's out for now.

ANNABEL

Uh-huh.

CHRIS

Once we've raised the money for our building, no problem. We can say whatever we want.

ANNABEL

Yeah.

CHRIS

I've got to go. Do you want to talk about this more later?

ANNABEL

No, I'm done talking.

SCENE 28

ALEX and ANNABEL at home in Montreal. ANNABEL has just returned from a meeting with her theatre company's board of directors.

ALEX

How'd it go tonight?

ANNABEL

I made another rash announcement to the board.

ALEX

Okay.

ANNABEL

I asked them to triple the travel budget.

ALEX

Why? Where are you going now?

ANNABEL

I told them you and I have to take our kids out of school for two months to drive across Canada.

ALEX

Um, when?

ANNABEL

ASAP.

ALEX

Why the urgency all of a sudden?

ANNABEL

Because if we wait I have a feeling it will be too late.

ALEX

Where do you want to go exactly?

ANNABEL

The oil sands.

ALEX

You want to drive to the oil sands?

ANNABEL

I want to experience the distance between here and the oil sands. I want to get up close to something that we talk about all the time but that we've never seen with our own eyes. I want to meet at least one human being in the Conservative Party face to face. I want the kids to realize what Canada is.

ALEX

That's a lot of hours in our car.

ANNABEL

We could get ... another type of vehicle.

Pause.

ALEX

What did the board say?

ANNABEL

What they always say – that I'm crazy.

ALEX

But they're not going to stop you.

ANNABEL

Well, that's one advantage of being president of your own board.

ALEX

Can we afford it?

ANNABEL

They said the most we can afford is four weeks. I might need to ask my dad for a loan until we get our grants this summer.

ALEX

What about *X-Men*?

ANNABEL

Who?

ALEX

That film I just got cast in? That's going to pay our kids' tuition next year?

ANNABEL

We'll have to work around that when you get your dates. Shall we talk with the kids and see what they think?

ALEX

Do we even need to?

The next morning.

ELLA and BEATRICE jump up and down, high-fiving and squealing with joy.

ELLA

We're missing four weeks of school?!

BEATRICE

Yes! Yes! Yes!

ANNABEL

Wait a second, you're not missing school. This will be your school.

BEATRICE

Like *School of Rock*!

ANNABEL

No, like School of ... Water, with your parents.

BEATRICE

You rock, Mama!

ALEX

And we'll have to keep up with schoolwork while we're on the road.

BEATRICE

Awww!

ELLA

On the road? We're driving?

ANNABEL

Yes.

ELLA

In our car?

ALEX

Uh, we're not sure yet.

ELLA

You're not sure. Which means maybe we'll get a bigger car!

BEATRICE

Yes!

ALEX

Guys, calm down. None of this is confirmed yet.

ELLA

Can we invite Hazel to come with us?

ANNABEL

No!

BEATRICE

Oh my God, HAZEL!

ALEX

Hold on a second ...!

ELLA

You always say we're much easier to handle when we have a friend over, Mummy!

BEATRICE

Yeah, we promise we'll work really, really hard learning about water! We never get to spend time with Hazel.

ANNABEL

We hardly have enough room in our car for us.

ELLA

But Daddy said we're getting a bigger car for the trip.

ALEX

I never said that.

ELLA

You said –

BEATRICE

I know! We could get a WINNEBAGO!

ELLA

Oh my God! A WINNEBAGO!

BEATRICE

A WINNEBAGO!

ELLA

Yes! Yes!

BEATRICE

Oh please, Daddy! Can we get a Winnebago? Please?

ANNABEL/ALEX

No.

ELLA

But why not? Then you'll save so much money because we don't have to stay in hotels.

BEATRICE

And we'll save so much water because ... because ... does a Winnebago have a toilet in it?

ELLA

Of course, Beatrice. A toilet, a shower, beds, TV!

BEATRICE

TV?!

ELLA

It's like Barbie and the Dreamhouse. You press a button and then – these, like, these hammocks come out of the top. And there would be, like –

ALEX and ANNABEL have given up stopping the juggernaut of their kids' fantasy.

BEATRICE

A Jacuzzi that came out too.

ELLA

And big closets with all our clothes in it. And a –

BEATRICE

And there's, like, a chandelier! And like ...!

ANNABEL

(*aside to ALEX*) For the record, I will never, ever, ever, live in a Winnebago with my children for any length of time, no matter how big it is.

ACT 2

SCENE 1

Day 1 of road trip. Montreal, April 2013. ALEX, ANNABEL, ELLA, and BEATRICE are hurriedly packing a Winnebago.

Sound of opening theme music to the CBC Radio show Q *with host JIAN GHOMESHI.*

JIAN GHOMESHI enters.

JIAN GHOMESHI

Well, hi there! Happy Friday! And in the great CBC tradition of tales of intrigue and adventure in the Canadian heartland, what say you to an idea like this: A new TV series about scientists, fresh water, and tensions in the halls of power? Think *Danger Bay* meets *The Nature of Things* meets question period. Maybe? Call me? Call it ELA: Experimental Lakes Area. Based on a true story.

ANNABEL

Alex? Where's the bag with the laundry detergent?!

SCHINDLER and DIANE enter.

JIAN GHOMESHI

It was about a month ago that our federal government shuttered the research facility in northwestern Ontario.

Well, this week the government of Ontario made a pledge to support the Experimental Lakes Area.

Enter KATHLEEN WYNNE.

JIAN GHOMESHI

On Wednesday morning, Ontario Premier Kathleen Wynne said –

KATHLEEN WYNNE

It's important to us, a government who believes in science and believes in evidence, to have this continue.

SCHINDLER

Premier Wynne's intervention is like a ray of sunshine in the Dark Ages that the Harper government has planned for Canada.

KATHLEEN WYNNE

Investing in science to help us understand and prevent pollution is a wise investment for the people of Ontario.

JIAN GHOMESHI

Enough said. But how will it be done? Ottawa still holds the cards and says that it still wants to lead the transfer to a new private-sector operator.

DIANE

Nothing happens until the government rubber-stamps this deal.

JIAN GHOMESHI

In a statement to Postmedia News, Fisheries and Oceans Minister Keith Ashfield, quote, "welcomed" Ontario's announcement but said it would be inappropriate to provide further comment.

SCHINDLER

Strange that Ottawa didn't want to comment.

DIANE

Why didn't the feds jump on this opportunity to express good faith about the transfer?

JIAN GHOMESHI

Call it ... a fluid situation. For now, looks like the pressure is on the prime minister. The Experimental Lakes Area may have a lifeline. We'll see what Ottawa will do. I'm Jian Ghomeshi. This is *Q*.

> *Sound of the* Q *theme music morphs into the theme music for CBC Radio's* As It Happens. *Meanwhile, ANNABEL and ALEX are helping ELLA and BEATRICE get ready to leave.*

ANNABEL

Okay, we've got to push back in five, everyone! Alex – I've got the girls; you get the booze bag!

ALEX

Beatrice, get out here and take care of Charlotte!

> *CAROL OFF, host of* As It Happens, *interviews GREG RICKFORD, then Conservative MP for Kenora.*

CAROL OFF

As you may have heard on this program yesterday, an agreement is now officially in place that will allow the Experimental Lakes Area to be transferred to the Winnipeg-based International Institute for Sustainable Development.

ANNABEL

Ella, did you pack your books? What about your toothbrushes, girls?

CAROL OFF

Greg Rickford is a member of the Harper government. He's the Member of Parliament for Kenora, the riding in which the ELA is located. We reached Greg Rickford today in Ottawa.

ALEX

I'm locking up the house now!

CAROL OFF

Mr. Rickford, this is good news!

GREG RICKFORD

Yeah it is good news, Carol. Obviously partly relieved and partly felt like – particularly for the benefit of my constituents – that they always understood that this has always been a priority, that we always wanted to find a third-party operator for the ELA.

ANNABEL

Come on! We're already an hour late!

CAROL OFF

Some of the scientists that we've spoken to – and I'm referring here to a group from Trent University doing work on silver nanoparticles – say it's too late ... that their work has already been interrupted.

GRANDPA enters with a bottle of champagne to say goodbye to the family.

ALEX

Girls, come say goodbye to Grandpa!

GREG RICKFORD

Carol, I can't speculate on that. I should remark, as a matter of fairness, that there have been a lot of comments from some folks on this throughout the debate. We know that some of them have been a distortion, but I have a reputation of taking the high road, Carol.

CAROL OFF

I appreciate that you wanted to take the high road on that, but the scientists told us that the buildings were being torn down.

ANNABEL

(*to ELLA and BEATRICE*) In! We gotta go!

CAROL OFF

And they said that their research was in jeopardy.

GREG RICKFORD

Well –

CAROL OFF

Were they lying?

GREG RICKFORD

– more recently built facilities –

CAROL OFF

Were they lying?

GREG RICKFORD

– are in excellent condition –

CAROL OFF

No, just – I wanna know – If there's wrong information, where did it come from? Were they deliberately misleading us then?

ANNABEL starts the Winnebago. ALEX, ANNABEL, ELLA, and BEATRICE wave goodbye to GRANDPA.

ANNABEL

I'm backing up now!

GREG RICKFORD

You know what, Carol? I'm not going to speculate or comment on that. When these kinds of issues arise, you have to be very clear on the *facts* and very clear on what's *going on.* My responsibility to my constituents – and in this case, the media – is to answer their questions with clear statements of fact. And that's what was going on.

Pause.

CAROL OFF

Mister Rickford, thank you.

GREG RICKFORD

Thank you for your time, Carol.

ALEX

Seatbelts on!

ANNABEL

Carol: Off!

SCENE 2

Later on Day 1, Highway 401, westbound. Inside the Winnebago, ANNABEL is driving, ELLA and BEATRICE are watching a movie, and ALEX is sleeping on the back bed.

Sound of Shrek 2 *movie soundtrack.*

ANNABEL

Okay, girls, time to turn off our screens!

BEATRICE

Wait! Can we just finish this ...?

ANNABEL uses a remote control to turn off the TV.

BEATRICE

Nooooooo!! Mama!!!

ANNABEL

Calm down.

BEATRICE

That was my favourite part!

ANNABEL

Yeah? How many times have you watched *Shrek 2*? (*pause*) Okay, first question of the trip, Bibi. What is ... a *watershed*?

Pause.

BEATRICE

It's a ... a water shed. A shed with water in it.

ANNABEL

Incorrect.

ELLA

Come on, Beatrice.

ANNABEL

A watershed is a system of interconnecting lakes, rivers, swamps – *any* bodies of water that all drain into the same basin.

ELLA

Like into the same pipe?

ANNABEL

No, well, we're talking about a natural system now. So one watershed is distinct from another watershed by the fact that they don't –

ANNABEL waits to see if ELLA and BEATRICE know the answer.

ANNABEL

– share the same basin! And the best description I've heard so far of how watersheds are defined in our natural landscape is, if you were to put two empty glasses side by side and pour water over the top of them, some of that water will pour into the glass on the right side, and some would pour into the glass on the left side.

BEATRICE

(*to ELLA*) I don't understand.

ANNABEL

So on this trip we are going to be travelling through many watersheds.

ALEX wakes up and goes to the toilet.

BEATRICE
But where are we *going* to again?

ANNABEL
Well, we're going to start in Stratford to pick up Hazel.

ELLA
Yes.

BEATRICE
No, but where are we going to *end up*?

ANNABEL
Well, our final destination is Fort McMurray, which is in –

ANNABEL waits to see if ELLA and BEATRICE know the answer.

ANNABEL
– northern Alberta.

ELLA
And what is the watershed there called?

ANNABEL
Good question, Ella. I actually have no idea. See how much we're going to learn on this trip?! Okay. Do either of you have a question to ask me?

BEATRICE
Are we gonna be polluting a lot on this trip, Mama?

ANNABEL
Yes, we are. But we're gonna offset our fuel consumption by planting trees.

ELLA
We're going to plant trees?

ANNABEL

No, we're not going to plant trees ourselves. We're going to pay so that other people can plant trees.

BEATRICE

Other people who?

ANNABEL

People whose expertise it is to plant trees while we employ our expertise making a play in our Winnebago.

She uses intercom system to talk with ALEX in the back.

ANNABEL

Actually – Alex sweetheart – can you Google how to do the carbon offsetting? We need to set that up. (*pause, then to ELLA and BEATRICE*) Any other questions?

BEATRICE

How many more minutes until we get to Stratford?

SCENE 3

Day 2, Stratford, Ontario. ELLA and BEATRICE squeal with joy when they see HAZEL. Meanwhile, DR. HANK VENEMA sends ANNABEL an online message, which she reads while LIISA, HAZEL's mother, dashes in and out of the house bringing loads of HAZEL's poorly packed luggage into the Winnebago.

ELLA/BEATRICE

HAZELLLLLLLL!!

ELLA and BEATRICE show HAZEL around the Winnebago.

LIISA

I packed some kids' Gravol. You know that Hazel tends to get carsick, eh?

ALEX

Hazel gets carsick?

HANK

(*online message*) Annabel. Hank Venema here. Sorry for the radio silence. I'd be happy to do an interview with you when you pass through Winnipeg. What are your dates? These are what work for me.

LIISA

And please ask her to be careful because she broke her tooth last week and we had a cap put on, but if she falls down it could come off.

ALEX

Okay, we'll do our best.

ANNABEL

Alex, when are you flying back for *X-Men* again?

ALEX

The twenty-fifth.

ANNABEL

Shit.

LIISA

I packed some arts-and-crafts stuff. She loves crafts.

ALEX

(*to ANNABEL*) What?

ANNABEL

We have to be in Winnipeg the next day.

ALEX

Why?

LIISA

These are some of her favourite DVDs. *The Sound of Music* really cheers her up.

ANNABEL

Liisa, if we don't leave in five minutes we're not going to make our ferry to Beausoleil Island.

LIISA

Okay, there's just a few more bags.

ANNABEL

(*to ALEX*) Hank Venema from the IISD just surfaced. We can interview him in Winnipeg on the twenty-sixth.

ALEX

My flight is from Thunder Bay though.

ANNABEL

We'll have to change that. I can't do that drive alone in one day.

LIISA

Can you play "Edelweiss" on the guitar, Alex?

ALEX

Just ask Hank to wait a few days.

LIISA

Chris sings "Edelweiss" to Hazel when she can't get to sleep.

ANNABEL

Just ask Jennifer Lawrence to wait a few days.

ALEX

(*to ANNABEL*) Excuse me?

ANNABEL

They can't shoot that scene without you.

LIISA

Oh my God, I almost forgot to give you her health card! Chris! Where is Hazel's health card?!

LIISA runs back into the house.

ANNABEL

You gave your unavailabilities. They changed the dates on you three times.

ALEX

And offered to pay all my flights home.

ANNABEL

Which is a drop in the bucket for a Hollywood budget.

ALEX

Are you out of your mind? You think the *X-Men* production is going to stop to accommodate the actor playing Quarantine Doctor Number 2?

ANNABEL

And what if I miss Hank?

ALEX

Well, you've been doing just fine without him so far.

ELLA and HAZEL enter.

ELLA

Can we go now, Mama?

HAZEL

Yeah! Let's get this show on the road, everyone!

LIISA enters with CHRIS.

LIISA

Thank God! We found it! Chris, should I go get her vaccine record too?

CHRIS

She'll be fine. Hazel, come give me a hug. I've got to get to rehearsal.

HAZEL

Bye, Dad. I love you.

CHRIS

I love you too. You've got your Gravol?

HAZEL

Yup.

ALEX

Okay in the car, everyone!

LIISA

Wait! Hazel!!

HAZEL

(*from inside*) Bye, Mum!

ANNABEL turns the key in the ignition.

ALEX

Wait a second! Where's Beatrice?

ANNABEL

Beatrice!

BEATRICE

(*offstage*) Coming!

ELLA

What the heck, Beatrice?!

BEATRICE

Sorry!

The Winnebago leaves.

SCENE 4

Day 5, Ontario Highway 17, westbound. ALEX drives the Winnebago while ELLA, BEATRICE, and HAZEL watch The Sound of Music.

ANNABEL writes an email.

ANNABEL

Dear Mr. Rickford,

I feel compelled to interview you because so far I have had trouble speaking with any Conservative MPs about the ELA transfer. Given that the ELA is in your riding of Kenora – where my family and I will be passing through next week – I feel you are uniquely poised to clarify certain questions I have.

ALEX

(*to a driver who has cut him off*) C'mon! Indicate, buddy!

ANNABEL

My interest in speaking with you is not about adding sensationalist fodder to the current daily news cycle. It's about ... breaking away from all the entrenched ideological warfare in our country to build new channels for dialogue in Canada. More concretely, I am gathering interview material for a play I am writing that will premiere in Toronto during the 2015 Pan Am Games.

Lighting change suggests a shift to later that day.

ANNABEL

(*looking at her smartphone*) OH MY GOD!

ALEX

Shit ... what?

ELLA

What? Did you see a moose, Mama?!

HAZEL

A moose! Where?!

ANNABEL

This is better than a moose, guys. We are going to do an interview with someone really important.

HAZEL

Stephen Harper?

ANNABEL

No, but someone who works closely with Stephen Harper: Greg Rickford!

ALEX

No way!

ANNABEL

Yes way, baby. I just got an email from his assistant. That was fast.

ELLA

Who's Greg Rickford?

ANNABEL

The MP who represents the riding where the ELA is located. Our first Conservative MP. We're going to interview a Conservative, ladies!

HAZEL

Awesome. Now we can finally get someone to admit that Stephen Harper *did* smoke crack cocaine.

ANNABEL

That's Rob Ford, Hazel.

HAZEL

No it's ... (*pause*) Oh yeah.

ELLA

But Mummy, I thought the ELA was saved by Ontario. So why do we still need to talk to Conservatives?

ANNABEL

Because even though Ontario has pledged money, it's not clear if Stephen Harper is going to let the transfer happen.

ELLA

I don't get this ELA story at all.

ANNABEL

The thing is, this isn't a Disney movie, darling.

ELLA

What is it then?

ANNABEL

It's ... it's ... the ELA has become, like, You-Know-Who in *Harry Potter*. If we say the name out loud it's like ... ushering in the Death Eaters.

HAZEL

Is Stephen Harper a Death Eater?

ANNABEL

No. No, darling, that's not what I'm saying. All I'm saying is, Greg Rickford has accepted our interview request, which means he's not afraid of us. Now we have to prepare to *not* be afraid of him.

ELLA

How are we going to do that?

BEATRICE

We have to think really good thoughts about Greg Richford.

HAZEL

*Rick*ford.

BEATRICE

That's how they make Patronuses to survive the Dementors in *Harry Potter*. They think really good thoughts.

ANNABEL

(*approving*) Think good thoughts. I think the kids are ready to start asking some questions.

SCENE 5

Day 8, Landslide Trailer Park, Sault Ste. Marie, Ontario. ELLA, BEATRICE, and HAZEL roast marshmallows at a campfire outside the Winnebago. ELLA records the conversation with ANNABEL's sound recording device.

ELLA

Okay so, Beatrice, explain to Hazel: What is exactly a watershed? Hazel, your marshmallow is on fire.

HAZEL

That's the way I like it – (*blowing it out*) burnt.

BEATRICE

A watershed is like two bowls – well actually, two lakes – that are sticked together. And then there's rain that falls. And if it's on the right side, it'll fall into the right, um, lake. If it's on the left side, it'll fall into the left lake.

ELLA

Okay, so what also is a watershed, is ... a change, kind of, that's gonna happen. So, the watershed –

HAZEL

Can you pass me the graham crackers?

ELLA

The watershed is the change that's gonna happen in the world. A change when we realize, finally, that we're ... that ... that we're always just polluting every single water body.

And nobody knows what's good anymore. And what's really bad for the water. Because our government is not ... our prime minister is not ... giving money to the ELA.

STEPHEN HARPER enters and sits down with the girls around the campfire. He roasts his own marshmallow.

HAZEL

And also Stephen Harper only really cares about money.

ELLA

Exactly.

BEATRICE

Hazel, no –

HAZEL

And also he doesn't want to pay money to the ELA 'cause he wants to keep the oil *source.*

ELLA

Very good, Hazel.

BEATRICE

No, Hazel. You don't understand. Stephen Harper, he really wants to help the ELA except it's way too expensive for him.

LLA

No, Beatrice, it's Justin Trudeau who wants to help the ELA.

EATRICE

No.

AZEL

Oh, Justin's so hot.

ELLA

Yeah, but Grandpa says he's just not ready.

BEATRICE

No, *Hazel*. Stephen Harper ... do you know how much cheques he's giving to people every day? Do you know what the debt of Canada is, Hazel?

HAZEL

Nope.

ELLA

Beatrice –

BEATRICE

Six thousand hundred billion dollars.

ELLA

Beatrice, the ELA costs only two thousand dollars. Compared to other companies, that's tiny.

HAZEL

Can you pass me the bag of marshmallows?

BEATRICE

But –

ELLA

So that is tiny compared to, like, other –

BEATRICE

Because once you said in an interview it cost ten million dollars for one scientist.

ELLA

I never said that, Beatrice.

BEATRICE

Yes you did!

ELLA

Beatrice, Mummy already explained that to you a lot.

BEATRICE

Okay, Ella! No fighting while it's on an interview, okay?

ANNABEL calls from inside the Winnebago.

ANNABEL

Girls, time for bed.

ELLA

It's not ... I'm not fighting ... It's just that you got it wrong. (*pause*) Okay so, Hazel?

HAZEL

Yeah?

ELLA

Do you get what a watershed is now?

HAZEL

Uh-huh.

ANNABEL enters.

ANNABEL

Come on, time for bed.

ELLA, BEATRICE, and HAZEL exit. STEPHEN HARPER starts to sing "With a Little Help from My Friends" by The Beatles. ANNABEL listens and hums along.

SCENE 6

MAUDE BARLOW enters.

MAUDE BARLOW

Think about what you're doing, Annabel. Every time they build a new pipeline, they reaffirm the idea that oil and gas should be our energy source for decades to come. We need an alternative narrative, and that alternative narrative has to be that we have got to stop *believing* in unlimited growth. The planet cannot sustain unlimited growth.

ANNABEL

But Maude that's a radical idea.

MAUDE BARLOW

It doesn't matter if it's radical. It's true! But that doesn't mean we can't develop, and it doesn't mean that anybody wants anybody to go back to living in caves.

ANNABEL

But right now most Canadians support Harper's economic –

MAUDE BARLOW

Most Canadians? Stephen Harper doesn't care about being prime minister for most of us – he wants to be prime minister of the 30 percent of people who would agree with the dismantling of protections for our fresh water. Right now most Canadians are struggling economically. Our personal debts are through the roof. But they get away with this ... this masquerade that they're the ones who know the economy. It makes me crazy!

ANNABEL

So what are *you* doing, Maude, to reach those people who believe that?

MAUDE BARLOW

I'm not doing *anything*. Hard core right-wing people don't want to be reached by me.

ANNABEL

But aren't those hard core right-wing people the ones you should be talking with if you really want to change the narrative?

MAUDE BARLOW

No.

ANNABEL

Why not?

MAUDE BARLOW

Because they're ideologues, Annabel! People who strongly believe that what they're doing is the only way it should be done.

ANNABEL

But by that definition you're an ideologue too, Maude, and you know as well as I do that the inevitable outcome of hard line ideology is violence. It's the end of dialogue. It's war.

MAUDE BARLOW

No, it's *revolution*.

SCENE 7

Day 16, Ontario Highway 11, westbound. Inside the Winnebago ALEX, ELLA, BEATRICE, and HAZEL are watching the CBC documentary film Tipping Point: The End of Oil, *written and directed by Tom Radford and Niobe Thompson, while ANNABEL drives.*

FIRST NATIONS ELDER

What you do with your money is your business. But when you spend your money in my territory, then that becomes my business.

FILM NARRATOR

For Chief Allan Adam and generations of his people, the land is life. Then one day the land changed. The world's largest corporations came looking for oil. Alberta's controversial oil sands – the biggest deposit of oil on the planet – are the largest source of imported oil to the U.S. But as the cost to the environment is becoming clearer, could new scientific evidence trigger a tipping point for the oil sands?

BEATRICE

Whoa.

ALEX pauses the video.

ELLA

Those are the oil sands?

ALEX

Okay so ...

ANNABEL

Did you hear that? "The biggest deposit of oil on the planet" is where we are heading right now, girls.

HAZEL/ELLA

Whoa.

ALEX

Okay, but –

ANNABEL

Incredible, eh?

HAZEL

Yeah! Because that's where the most dinosaurs died!

BEATRICE

And also, I think maybe the oil got exploded so it killed all the dinosaurs!

ALEX

Okay, hold on a second. First of all, why do you think they interviewed First Nations people for this film?

Pause.

BEATRICE

Because ... First Nations people were the first people on our land so ... they know the most about ... land ... stuff.

ALEX

Yes. Their families go back many, many generations, before we started drilling for oil, before we started zooming around in Winnebagos. So they have an actual relationship to the land that is deeper than ours, which we, unfortunately, are not going to learn about on this trip.

ELLA

Why not?

ALEX

Because Mummy says we don't have time.

ANNABEL

Oh give me a break. They're watching the movie.

ALEX

They're watching a movie while, outside that window, we are passing by hundreds of opportunities to interview live First Nations people on their own land.

ELLA

Why can't we stop, Mama?

ANNABEL

Ask my board of directors.

BEATRICE

But you're the president of your board.

ANNABEL

Yes, but I'm not the treasurer. You guys have ten kilometres to finish the movie before we stop for lunch.

ALEX reluctantly puts the movie on again.

SCENE 8

Day 19, gas stop at Shabaqua Corners, Ontario. ANNABEL fills up the Winnebago.

Sound of ANNABEL's smartphone ringing.

ANNABEL

(*answering phone*) Hey.

CHRIS

I called, like, three times this morning.

ANNABEL

I know. Alex is back in Montreal filming and we're on the road all day. Can I call you tonight when we get to our trailer park?

CHRIS

Um, I think I have to tell you this now.

ANNABEL notices ELLA and HAZEL walking into the gas station.

ANNABEL

(*to CHRIS on phone*) Okay, just a sec – (*to ELLA and HAZEL*) No chips! No chocolate bars! No ice cream! No gum! Something with some *numbers* in the bottom part of the nutritional facts box!

Sound of car horn.

ANNABEL

Ella, can you tell that lovely lady in the pickup truck that we will move in five minutes?

ELLA

Why can't Beatrice do it? ... I always –

ANNABEL

Ella!

ELLA exits.

ANNABEL

(*to CHRIS*) Sorry, go ahead.

CHRIS

Don Shipley called me this morning.

ANNABEL

Yeah?

CHRIS

He got a call yesterday from the Ministry of Culture in Ontario.

ANNABEL

Uh-huh.

CHRIS

Who in turn had received a call from the federal government.

ANNABEL

Okay.

CHRIS

They were asking about your play.

ANNABEL

About my play?

CHRIS

They asked him some questions about the content of the play and about why he is programming us at the Pan Am Games.

HAZEL enters from gas station.

HAZEL

Did you know that beef jerky has a lot of potassium in it, Annabel?

ANNABEL

(*to HAZEL*) Just a second, darling.

HAZEL exits.

ANNABEL

(*to CHRIS*) Chris, I'm not wanting to compute here.

CHRIS

They found out somehow that we're doing a play about the ELA, and apparently they're not happy about it.

ANNABEL

So what are we going to do?

CHRIS

I don't know. I've got a one-point-five million dollar application in to the feds right now for our new building.

ANNABEL

Oh my God.

CHRIS

It's okay. We're okay for now. Don stood up for us.

ANNABEL

For *now*?

CHRIS

I know, it's crazy. I'm totally freaked out.

ANNABEL

Who even ...?!

CHRIS

It sounds like someone from Rickford's office got in touch with someone to ask them why they were funding this project and the shit hit the fan. It must be because of the email you sent him.

ANNABEL

But I'm trying to reach out to them, that's the whole point.

CHRIS

It doesn't matter. The ELA story is too volatile.

Sound of car horn.

ELLA and HAZEL enter from the gas station.

ELLA

Mama, we have to go.

ANNABEL

(*to CHRIS*) I have to go.

CHRIS

Just ... maybe let me know before you communicate with anyone else.

ANNABEL

We're interviewing Rickford in two days, Chris.

CHRIS

Okay, but we got to talk before about how you approach that –

ANNABEL

Excuse me?

CHRIS

I don't want to cave, Annabel, but maybe we have to be a bit more strategic about how we talk about the show from now on.

ANNABEL

Chris, I am going to interview Greg Rickford on Monday morning. I am going to be very open, very polite, and I am going to listen very carefully to what he has to say. But I am not going to mince my words about what this play is about. I have come too far for that.

Sound of car horn. ANNABEL hangs up on CHRIS and yells outside to the lady in the pickup.

ANNABEL

I know you want me to move, lady, but my vehicle is a lot bigger than yours!

ANNABEL gets back in Winnebago.

ELLA

Where's Beatrice?

ANNABEL

BEATRICE!

ELLA

BEATRICE?

BEATRICE enters.

ELLA

I was walking Charlotte.

SCENE 9

Day 21, Anicinabe RV Park and Campground, Kenora, Ontario. ALEX, ELLA, BEATRICE, and HAZEL are preparing breakfast on an outdoor grill at a campsite.

ELLA

That's my piece of bacon, Beatrice.

BEATRICE

No, I put it on the grill.

ELLA

But I said! I said! Daddy, you promised I would get the first piece!

ELLA and BEATRICE physically fight over the bacon. ANNABEL suddenly opens the door of the Winnebago.

ALEX

Girls! Sit!

ELLA and BEATRICE sit.

ALEX

Stay!

ANNABEL

Okay, everyone, be quiet for a second! Shut the hell up!

ELLA and BEATRICE are quiet.

Sound of the radio news playing from the cab of the Winnebago.

CANADIAN PRESS REPORTER

The Right Honourable Prime Minister Stephen Harper announced this morning that he was naming Kenora MP Greg Rickford to his Cabinet.

GREG RICKFORD

Today, I was honoured to be asked to serve Canadians as the Minister of State for Science and Technology. A critical part of our government's Economic Action Plan is advancing science and research to secure Canada's long-term prosperity.

ANNABEL

Oh my God.

ANNABEL checks her smartphone for online news about the announcement. ELLA edges closer to the grill to grab bacon.

ALEX

Looks like we're scooping an interview with a newly appointed cabinet minister, girls.

BEATRICE

Give me that, ELLA!

ELLA

Beatrice!

ALEX

Stop! Did you even hear what they just said on the news?

HAZEL

Yes. That's terrible. That guy is against the ELA and now he got promoted?

ALEX

Very good, Hazel. At least somebody's thinking about more than her next piece of pork.

ALEX takes bacon from ELLA and gives it to HAZEL.

ELLA

What? No fair!

ALEX

So in two hours, Hazel, we are going to be interviewing "that guy" – so we better get ready to ask some good questions.

HAZEL

Oh, I'll be ready.

ANNABEL

Actually we won't be interviewing that guy.

ALEX

What?

ANNABEL passes smartphone to ALEX so he can read email from Greg Rickford's communications assistant.

ALEX

(*reading email aloud*) Good morning, Annabel, my apologies for the last-minute change of plan; however, with Minister Rickford's recent appointment to Cabinet, he will be leaving the riding immediately to travel to Ottawa and will therefore not be available for an interview with you and your family later today.

ANNABEL

Fuckhead. I hate you.

BEATRICE

(*to ALEX*) You said we're not supposed to say "hate," Daddy –

ANNABEL

I loathe you. I despise you.

ALEX

Annabel –

ANNABEL

No. I just drove a thousand kilometres to be told "my apologies for the last-minute ... " Fuck you.

HAZEL

My mum swears a lot when she's mad too.

ALEX

We still have an appointment. Let's just go to his office –

ANNABEL

Agh!

ELLA

I promise I won't ask for a treat today, Mama.

BEATRICE

Yeah, we'll be really, really good today, Mama. Let's just go to Richford's office.

HAZEL

*Rick*ford.

ANNABEL

I don't want to go anymore.

ALEX

Oh come on.

ANNABEL

No, I don't want to go anywhere. I want some bacon. (*pause*) I WANT SOME BACON!

ANNABEL sees the dog eating the leftover bacon.

ANNABEL

CHARLOTTE! BAD DOG, CHARLOTTE!

ALEX ushers ELLA, BEATRICE, and HAZEL into the Winnebago.

ALEX

Go put your watershed T-shirts on for our interview, girls.

BEATRICE

But mine's got ketchup on it.

ALEX

Just put it on.

ELLA, BEATRICE, and HAZEL exit.

ANNABEL

I know what you're going to say, but you know what? I want them to see bad behaviour. I want them to experience that because then maybe they'll say to themselves, "I don't like to see my mother scream for bacon. Maybe I should stop behaving like that."

ALEX

That's really excellent reasoning, Annabel.

ANNABEL

I'm not used to this doing research with three kids. So – yes, I'm having a difficult time trying to make this trip be worthwhile, you know? And then I have Chris calling from Stratford telling me to be *strategic* about working on this play. I mean, I'm not even ... *am I even working on this play*?! No! I'm taking care of three children who think they're on vacation. I'm working out itineraries and calling ahead to book campsites. And trying to work around my husband's *X-Men* schedule.

ALEX

No, you're *collecting*. You're collecting now. You *are* working. And stop qualifying it as if nobody else is working. You're disappointed about Rickford. I understand. But we've done more than fifty interviews with you since we left Stratford. We are not the obstacle here. We are with you, even though you probably wish you were alone.

ELLA, BEATRICE, and HAZEL eavesdrop from inside the Winnebago.

ANNABEL

All I'm doing is explaining why this is a difficult proposition for me, okay? I mean, I don't even have an office to retreat to, to make phone calls. I'll be sitting on that pathetic excuse for a bed back there, talking to someone, and Hazel will, like, leap over me and, like –

ALEX

I know.

ANNABEL

– knock out my power cord and then the girls don't even have the sense to not ask me a question when I'm on the telephone about where they can find a *fork*. We've been living in this fucking shoebox for three weeks. Is it possible they don't know where the cutlery is? I mean, what does that even say about us as parents?

ALEX

It says that we're still learning.

ANNABEL

We're *failing*. We're fucking ... I mean, how much money have we spent on gas? How many Styrofoam coffee cups have we consumed? How much fucking beef jerky has Hazel eaten? What are we ...? What are we doing here?

ALEX

We're recording what is happening here.

ANNABEL

To what end? I've just been told that "this play" might never see the light of day. So – to what fucking end?

HAZEL

What did she say?

BEATRICE

She said, "To what f-word end"?

HAZEL

And what did he say?

ELLA

Nothing.

SCENE 10

Day 23, Saskatchewan Highway 16, northwest bound. Late evening. ANNABEL is driving. ALEX sits on the sofa. ELLA, BEATRICE, and HAZEL look out the back window of the Winnebago.

ELLA

Yeah but if you take away all the stars then there's nothing.

HAZEL

But you can't take away all the stars because it goes on forever.

BEATRICE

Yeah, but it can't go on forever.

HAZEL

Yeah, it can.

ELLA

What if you just suck up the whole earth, all of the planets –

BEATRICE

You can't!

ELLA

No no no, you suck up all the space ... The space that the earth was in ... Now there is nothing, you see?

BEATRICE

There's no space in nothing – there's nothing.

ELLA

Nothing.

BEATRICE

But that's what I mean. There must be an end of space. I mean, it can't just –

HAZEL

But there is no end of space!

ELLA

Mummy, is it true that there's no end of space?

ANNABEL

It is true, especially in Saskatchewan.

ELLA

Is space nothing?

BEATRICE

No, there's stuff in space. There's ... meteorites.

ELLA

What if you suck up all the stars and all the meteorites?

BEATRICE

Yeah but there's still space! Space is something.

HAZEL

If you're dead, you're nothing.

ALEX

Are you guys torturing us because we didn't let you watch a movie?

ANNABEL

They can't do anything if they're not watching a movie. They don't even exist without movies. Nothing exists without movies.

ELLA

I know! I know what nothing is! It's – nothing.

HAZEL

But how do you get to nothing? That's our question.

ELLA

Wait, wait, wait! Before all the planets, before all the meteors –

HAZEL

But there *is* no before.

ELLA

Before all the space ... yes, there is!

HAZEL

You mean when the earth started?

ELLA

No! Before the earth started!

HAZEL

Before every planet in the world started?

ELLA

Yeah, there was nothing!

HAZEL

But scientists haven't even figured that out yet.

ELLA

Yeah well, so what? I just figured that out!

HAZEL

But how do you *know*?

ELLA

Because I just figured it out by thinking about it myself. What *would* there be before all those planets? Nothing! That's what I'm telling you!

BEATRICE

There was God.

ELLA

There is no God, Beatrice.

BEATRICE

God.

ANNABEL

Okay, you little Nietzsches, it's time to *do* something. You said when we started this trip you were going to choreograph a dance about water. So get to it.

HAZEL

I feel sick.

ANNABEL

Go take some Gravol.

HAZEL

I already did.

ANNABEL

Take some more.

HAZEL

I don't think I'm allowed –

ANNABEL

Don't start with me, Hazel. I have a headache. Go make up a dance now or nobody gets to swim in the pool when we get to the hotel in Fort McMurray.

Lights indicate scene transition.

Day 25, Radisson Hotel, Fort McMurray, Alberta. ELLA, BEATRICE, HAZEL, and ALEX swim in the hotel pool. Their swim is like a dance about water.

SCENE 11

Later on Day 25. Fort McMurray. ANNABEL, ALEX, HAZEL, ELLA, and BEATRICE are squeezed into a booth at a sushi restaurant.

WAITRESS

Will that be all?

HAZEL

Can I get some seaweed salad too?

ALEX

You ordered twenty-four pieces of sushi already, Hazel.

HAZEL

I know but I'm craving seaweed salad.

ELLA

Yuck.

ANNABEL

If you promise to do an interview after we order, we'll get some seaweed salad.

HAZEL

I promise.

WAITRESS

And one seaweed salad.

ANNABEL

If you could bring the bottle of sake right away, that would be wonderful.

WAITRESS

You got it.

The WAITRESS exits.

ANNABEL

Okay, we didn't come all the way to Fort McMurray to eat Japanese food. Here's the recorder.

BEATRICE

But who do we interview?

ANNABEL

Follow your gut. Find someone in this place who intrigues you –

ALEX

And who looks friendly –

ANNABEL

And remember to let *them* do the talking –

ALEX

And be polite –

BEATRICE

Okay, okay. (*to HAZEL and ELLA*) Come on.

ELLA

Can you please come with us, Mama?

ANNABEL

No.

ELLA, BEATRICE, and HAZEL approach a couple in their early thirties.

HAZEL

Um, do you mind if we ask you a couple of questions – about water?

MALE SUSHI DINER

About water?

HAZEL

Yeah.

ELLA

Because our parents are doing a documentary about water. Well, we are too.

MALE SUSHI DINER

Okay, sure.

BEATRICE

Uh, this is – oh, thank you very much by the way.

MALE SUSHI DINER

You're very welcome.

BEATRICE

So this is Ella. This is Hazel. And this is – (*giggles*) I'm Beatrice.

MALE SUSHI DINER

So what's your question?

Pause.

ELLA

You can ask the first question if you like, Beatrice.

BEATRICE

Thank you, Ella. Uh, what does the word "watershed" mean to you?

MALE SUSHI DINER

Watershed.

FEMALE SUSHI DINER

(*French Canadian accent*) Water ... shed.

Pause.

HAZEL

Do you ... do you know what a watershed is?

FEMALE SUSHI DINER

Um, no, actually.

HAZEL

You don't? Well, that's okay, don't feel bad.

FEMALE SUSHI DINER

Thank you, I won't.

HAZEL

Um, a watershed is a body of water –

ELLA

It's, like, *lots* of bodies of water.

HAZEL

Yeah, all connecting to each other.

ELLA

Yeah, like, a creek that connects to, like, rivers that connect to, like –

BEATRICE

Swamps.

HAZEL

Or wetlands.

MALE SUSHI DINER

Okay, a water system.

FEMALE SUSHI DINER

A system of waterways.

BEATRICE

Yeah, exactly. Very good! (*pause, then to ELLA*) So do you wanna ask a question, Ella?

ELLA

Thank you, Beatrice. Do ... do you think that, in your opinion, the water is much more polluted than before?

MALE SUSHI DINER

Most definitely.

ELLA

Yeah?

BEATRICE

Okay.

ELLA

Do you think that oil spills are bad for the lakes?

MALE SUSHI DINER

(*chuckling*) Yes, most definitely.

ELLA

Yeah. Okay, very good.

Pause.

BEATRICE

(*to HAZEL*) Okay. So do you have a follow-up question to that, Hazel?

HAZEL

Do you know about the *tar sands*?

MALE SUSHI DINER

I've heard of them.

ELLA

Did you ever *see* the tar sands?

MALE SUSHI DINER

Yep, I work in 'em.

ELLA, BEATRICE, and HAZEL gasp.

ELLA

Wow!

HAZEL

(*excited*) Really?! You work in the tar sands?!

BEATRICE/ELLA

Shhh!

HAZEL runs away to tell ALEX and ANNABEL the news.

HAZEL

Annabel! Alex! That guy works in the tar sands!

ANNABEL

That's amazing.

ALEX

Get back over there, Hazel.

ANNABEL

(*drinking sake*) Go ask him some tough questions about that.

ELLA

So, um, what type of work do you do in the tar sands?

HAZEL returns to continue interview with ELLA and BEATRICE.

MALE SUSHI DINER

I drill. I drill into the tar sands. And I see ... what material is at what levels.

ELLA

So, um, so what is it like to ... to ... drill there?

BEATRICE

In the tar sands?

MALE SUSHI DINER

Uh, it's dirty. It's dusty. And ... there's a lot of land disturbance. And ... water-resource disturbance. There's no doubt that it's all of the above. But at the same time, we must, um, we must get our resources one way or the other. And we have to keep drilling.

HAZEL

Have you ever seen a tailing pool?

MALE SUSHI DINER

A tailing *pond*? Yep.

ELLA

Okay.

HAZEL

Have you ever dipped your hand in a tailing pool?

MALE SUSHI DINER

Um, no. I don't actually go near the water.

HAZEL

Okay, we saw a video with, like, a hand going into a tailing pool –

BEATRICE

Pond.

ELLA

Yeah.

HAZEL

And it literally looks like mud, like, mud in, like –

MALE SUSHI DINER

Dirty, rusty mud?

BEATRICE

Yeah, it's really dirty.

MALE SUSHI DINER

Yeah, so ... But with all the negativity that does surround the oil sands, there is a lot of money invested into monitoring water systems here in Fort McMurray, did you know that?

BEATRICE

Um, no.

MALE SUSHI DINER

See, I don't only drill for exploratory purposes but we drill to *monitor* different water systems to see if our mining is influencing the water. So we do hurt the environment, but at the same time we closely *monitor* it to see how much we are hurting it and to see ... if we can mitigate any of these problems.

Pause.

BEATRICE

Yeah, it's true. It's true.

Long pause.

FEMALE SUSHI DINER

Where are you girls from anyway?

ELLA

We're from Montreal. (*referring to HAZEL*) She's from Toronto.

FEMALE SUSHI DINER

Alors vous parlez français?

ELLA/BEATRICE

Oui!

ELLA

Nous autres, oui. Mais pas elle.

HAZEL

French is my worst subject in school.

FEMALE SUSHI DINER

And did you fly out here to Fort Mac?

HAZEL

No, we drove in a gas-guzzling Winnebago.

FEMALE SUSHI DINER

Okay, so you saw a lot of empty land, right?

BEATRICE

Yeah, which our parents made us look at instead of watching movies.

FEMALE SUSHI DINER

Well, I'm glad they did because – this oil patch up here – people say it's big, and it is. But if you've driven across this country, you know that it's still a teeny tiny fraction of the land of Canada – *teeny* – and judging by how many people work in this town who used to live in Ontario and Quebec, it kinda feels like we need this place really badly right now.

SCENE 12

Day 27, inside the Winnebago. ANNABEL coaches ELLA, BEATRICE, and HAZEL while ALEX drives.

ANNABEL

That was a pretty impressive interview, girls. You asked clear, simple questions. You were polite. You didn't reveal a bias in your questions, until Hazel –

HAZEL

I know, I know ... the tailing pools.

BEATRICE/ELLA

Ponds.

ANNABEL

Remember, the experienced documentary researcher does not lead her subject to confirm a preconceived thesis, she –

BEATRICE/ELLA/HAZEL

Stays open.

ANNABEL

Stays open, that's right. To allow the subject to reveal his or her unique point of view. Now, everything we have heard about ... not the *tar* sands –

BEATRICE/ELLA/HAZEL

The *oil* sands.

ANNABEL

About the oil sands is at best second-hand information. We are now going to see them firsthand. And we're going to have an opportunity to ask a local pilot about their impact on the watershed. Now I want you to remember that no matter what you see up there in that plane this morning, you are weighing that truth against many of the highly intensive fossil-fuel burning lifestyle choices you make every day.

ELLA

That you make. We're minors.

ANNABEL

That's true. And yet, you were pretty psyched about this Winnebago.

BEATRICE

But you're the one who caved to our demands.

HAZEL

And I already told my dad I'd be willing to live like the Mennonites.

ANNABEL

Okay, okay ... fine. Our adult choices. But nevertheless ... (*to HAZEL*) hard to give up all the creature comforts, Miss Beef Jerky. And are we really ready to risk sabotaging our economy because we *think* the environment is suffering so much? No. We need to be sure. So no letting this guy off the hook like you did with that sushi guy. You have to have an *endgame*. What are you spending on environmental monitoring? How much pollution can we tolerate? Who exactly is doing the studies? What are the alternatives? When is enough, enough?

ALEX parks the Winnebago.

ALEX

Okay, we have to go, Annabel.

ANNABEL

We have to go. Any questions?

BEATRICE

Why can't we just walk around the oil sands?

HAZEL

Yeah, then I wouldn't get plane sick.

ANNABEL

They're too big to see from the ground.

ELLA

But can't we land inside them?

ANNABEL

No, there's no runway.

BEATRICE

Will we really be able to get the facts from the sky, Mama?

ANNABEL

I don't know, darling. But it's too late to change the game plan now.

SCENE 13

Later on Day 27, McMurray Aviation Inc. ALEX, ANNABEL, ELLA, BEATRICE, HAZEL, and OIL SANDS TOUR PILOT are squeezed into the cockpit of a Cessna airplane on the runway.

OIL SANDS TOUR PILOT

Okay, you'll all be sporting your own headsets for the trip this morning. Take a moment to set the volume levels.

ELLA squeals. The sound is amplified by the headset microphone.

ANNABEL

No screaming into the headsets!

ALEX

Jesus Christ, Ella!

ELLA

Sorry.

ANNABEL

Just calm down, everyone. We've all been in an airplane before.

HAZEL

But this one's so small.

Sound of engines as the plane begins to taxi down the runway.

BEATRICE

This is so cool!

HAZEL

I know!

ELLA

I'm really scared, Mama!

OIL SANDS TOUR PILOT

Tower – I'm doing Test 7335874 for oil sands tour.

RADIO TOWER

This is a confirmation.

OIL SANDS TOUR PILOT

Two-zero-four-six. Copy that.

ELLA

Are our ears going to pop, Daddy?

ALEX

I don't know.

HAZEL

Are you okay? Is everybody okay?

BEATRICE

Yeah.

HAZEL

Okay, good.

ELLA

Not really. I'm really scared.

ANNABEL

Don't be scared.

BEATRICE

We can do this! This is so fun!

ANNABEL

Okay so, girls, this is our opportunity to see what we've been talking about for so long, right?

ELLA/BEATRICE

Yeah!

HAZEL

Oh my God, I feel sick.

ANNABEL

Already? We're still on the ground.

HAZEL

I don't know ... I guess I feel fine.

OIL SANDS TOUR PILOT

We're almost ready to take off, guys. It's going to get real loud in a moment. But it's gonna be over real fast.

ALEX

Okay, guys!

HAZEL/ELLA/BEATRICE

Okay!

OIL SANDS TOUR PILOT

Right now, we're a go –

BEATRICE

We're going, Hazel!

HAZEL

Oh my God! Oh my God!

OIL SANDS TOUR PILOT

All right, guys. Here we go.

ALEX

Here we go!

ELLA puts her hands over her eyes.

ANNABEL

It's okay, Ella. You're going to be fine. Hold my hand.

Sound of plane engines picking up speed and then taking off.

HAZEL

We're off the ground! We're off the ground!

BEATRICE

I know! Wow!

ELLA

It's scary!

ANNABEL

Open your eyes. Look outside, Ella. Look outside. It's actually really beautiful. (*pause*) Okay, guys. Fire away.

ELLA

(*to OIL SANDS TOUR PILOT*) Um, is that an oil sand there?

ANNABEL

These are *all* the oil sands, darling. Ask another question.

ELLA

Um, which oil sand exactly is that?

OIL SANDS TOUR PILOT

To our right we see one of Suncor's mines. Over there on the left a little farther ahead is one of Syncrude's.

ELLA

Can you please tell me what kind of monitoring is done here on the watersheds to ... to –

BEATRICE

To mitigate –

ELLA

To *mitigate* the effects of the oil mining?

HAZEL starts to breathe deeply. ELLA continues to spew questions, with help from BEATRICE. In response to their questions, the OIL SANDS TOUR PILOT spews long sequences of numbers as if communicating location data to the control tower.

OIL SANDS TOUR PILOT

7750231.

ELLA

May I ask how we can trust that the results of those studies aren't being ... being –

BEATRICE

Being *influenced.*

ELLA

Being influenced by the people who are funding them?

OIL SANDS TOUR PILOT

49011722.

ANNABEL

Stay open! Stay open!

HAZEL is suddenly gasping for air.

ANNABEL

What's the matter, sweetheart? What's the matter?

HAZEL

I feel, like, really sick –

ELLA

Are you okay, Hazel?

ANNABEL

Just keep going, Ella. I'll take care of Hazel.

ELLA

Have you heard of the Experimental Lakes Area? Did you know that the ELA is the only site in the world where scientific research can … can?

BEATRICE

Can *objectively measure* –

OIL SANDS TOUR PILOT

1509934.

ELLA

Can objectively measure the long-term impact of oil sands mining and fossil-fuel consumption on fresh water?

ANNABEL

Just drink this, Hazel. You'll be fine. Alex! Are you filming?

ALEX

I'm trying!

HAZEL starts moaning.

ANNABEL

Keep going, Ella!

OIL SANDS TOUR PILOT

36701221.

ELLA

Did you know that the ELA is really cheap to operate?

OIL SANDS TOUR PILOT

726408.

HAZEL starts screaming with pain. HAZEL and ANNABEL's text begins to overlap with ELLA's question to OIL SANDS TOUR PILOT.

ELLA

What would you say if I told you that .0001 percent of total oil sands capital spending this year could keep the ELA platform running for twenty years?

ANNABEL

Hazel? Do you need more water?

HAZEL

No! No more water! Everybody, please stop talking! I can't breathe!

ANNABEL

Okay, Hazel. I want you to breathe consciously. Breathe in – One, two, three –

ANNABEL inhales deeply. HAZEL vomits suddenly.

ELLA

Oh my God! Oh my God, that is so disgusting! Hazel puked, Daddy! Daddy! Hazel puked!

ANNABEL

Ella, just keep going! You are doing *so well*!

HAZEL vomits again.

ALEX

Annabel, we have to go back.

ANNABEL

No, let her finish! Go, Ella!

ELLA

What do you personally envision for the future of the oil sands?

ALEX

We have to go back. (*to OIL SANDS TOUR PILOT*) Can you please turn this plane around?

HAZEL vomits again.

ALEX

One of our kids is very, very sick!

ANNABEL

She's not our kid.

ALEX

That's enough, Annabel!

HAZEL

Please stop talking, everyone! Your voices are all making me sick!

OIL SANDS TOUR PILOT

This is 7335874 requesting a landing.

ANNABEL

Just finish what you were saying, Ella. He didn't hear you, Ella! He didn't hear you!

Sound of the plane descending rapidly for landing. Sound of a big splash, as if plane has landed into water or oil.

SCENE 14

Day 30, Fort Garry Hotel, Winnipeg. ANNABEL and DR. HANK VENEMA stand together at the bar at the Fort Garry Hotel as in Act 1, but this time the scene feels like it is playing out as a dream.

HANK

So water is confronting us because water is the most disruptive agent of natural resource exploitation.

ANNABEL

Hank? Are you okay?

HANK laughs sadly.

ANNABEL

Mike told me the transfer has been hard for your team at the IISD. How are you doing with it?

HANK

You know, surviving.

ANNABEL

Yeah.

HANK

I need some time off.

ANNABEL

Are you gonna take it?

The WAITRESS enters.

HANK

I'm gonna have to 'cause I'm about to, you know, blow a fuse here.

ANNABEL

(*to WAITRESS*) Can I get a glass of water, please?

The WAITRESS ignores her.

ANNABEL

How many of you are working on the transfer? Like, it's a small team, right?

HANK

Yeah, it's about four of us at IISD working on this.

ANNABEL

Oh my God. That's the same size as my theatre company. I can't even imagine what you're going through institutionally to deal with this.

HANK

Yeah, you know it's, um ... we're dealing with some pretty uncreative people around the table right now. You know, DFO just wants to hand us the keys and run away – to save a few bucks. And Ontario is, like, reluctant to accept what they fear is more federal downloading, which they will have to pay for.

ANNABEL

No one wants to get stuck paying the tab. Even if it's to do the right thing.

HANK

No, in times of deep turbulence and uncertainty, everyone in government has retreated into pretty one-dimensional thinking. And the implications of that percolate, and we start to lose the assets that we will need for our eventual revitalization – well-resourced science, profound science in the public interest, that looks at the issues on a scale that matters, which is the ecosystem.

ANNABEL

Are you guys going to be able to run ELA with that vision now that you have to raise money from the private sector?

HANK

Well, you know what? No, not if the Canadian corporate and private sector doesn't smarten up, actually, and recognize this is part of the brand we wanna have. This is the crown jewel of Canada's brand. ELA is a totally visible demonstration of commitment to water stewardship. Like absolutely cutting-edge, creative, world-class water stewardship. If they don't see enough value to invest in that, I don't know what the private sector values anymore. (*pause*) Anyway, that's the idea we have to sell now. Wish us luck.

SCENE 15

Day 33, Ontario Highway 17, eastbound. Night. ANNABEL wakes up as if from a nightmare. ALEX is driving the Winnebago. ELLA, BEATRICE, and HAZEL are watching Star Wars.

ANNABEL

Where are we?

ALEX

Just crossed the Manitoba–Ontario border. Almost back on the Shield.

ANNABEL

What are you watching?

BEATRICE

Star Wars.

ANNABEL joins ALEX at the front of the Winnebago.

ANNABEL

Was this whole trip a complete disaster?

ALEX

I don't know yet.

SCENE 16

Brome Lake, Quebec. A morning in June. ANNABEL and GRANDPA are sitting on the lakeside patio of GRANDPA's cottage. GRANDPA is weak from chemotherapy treatment.

ANNABEL

I want to ask you a question about the future of capitalism.

GRANDPA

Uh-huh.

ANNABEL

And I wonder if our point of departure could be that we both agree that this is the best system to promote material well-being in the world.

GRANDPA

Yes.

ANNABEL

So that we don't take this as a debate.

Pause.

GRANDPA

So what's your question?

ANNABEL

My question is ... capitalism is an economic system, right?

GRANDPA

Yes.

ANNABEL

And I just wonder if you feel that the capitalist values you've always talked about –

GRANDPA

Yep.

ANNABEL

Are those values good for society outside of the economic sphere?

GRANDPA

That's a really good question. (*pause*) I don't know, and I don't know if there's any system that is good for society as a whole.

ANNABEL

Because my feeling today is that a lot of our decision-making is guided by economic values even though a lot of the decisions we are making –

GRANDPA

Sure.

ANNABEL

– are not economic in their nature. So when capitalism is applied, for example, to journalism, to ... what sells newspapers. It's not –

GRANDPA

It's bad news.

ANNABEL

It's bad news, and it's also bad journalism.

GRANDPA laughs.

ANNABEL

It's appealing to our ideology instead of our sense of what's really happening in the world.

GRANDPA

But it's – it's not black and white, darling. There's still all kinds of good journalists around that really do add value. You know, if you read serious press – you know, the *New York Times*, *Wall Street Journal*, *Financial Times*, etcetera, etcetera –

ANNABEL

Are there enough good journalists though?

GRANDPA

But all generations have worried about this, Annabel. You know, "We're slipping down a slope now and it's not as good as it was." And yet if you look at, sort of, the health of people, the affluence of people, etcetera, etcetera, we are making some progress. I mean ... we are. Certain groups are being left out, for sure –

ANNABEL

Certain groups? Dad, there's actually, like, fundamentally right now a tiny percentage of people who have a huge percentage of the wealth ... and I'm sorry but I don't have the stats with me here, but –

GRANDPA

But like all free-market cycles, darling, that will change. Maybe it'll take a financial crisis until we work through this inequality. But all systems have to go through these things to learn from the mistakes that have been made. And what

I'm saying is that I have more confidence in free markets to address that need for change than any other system.

ANNABEL

But you refer to a future that is heading back towards crisis, yet you say you're hopeful at the same time. How do you ...? I don't understand –

GRANDPA

I'm hopeful because I still think it is the best system. And I think, like ... *life*, you know? You have to go through a crisis to change. I mean, life's not straight up. The charts don't, you know – (*pause*) They don't always go straight up.

ANNABEL

No ...

GRANDPA

But ... but ... that *bottom* is where opportunities arise. Everybody in 2008 – I remember saying to everybody, "We should be buying stocks now." I mean, people said, "Are you crazy?"

ANNABEL

But you did it anyways.

GRANDPA

But I mean, they were unbelievably cheap at that point. And look at where the markets are now.

Pause.

ANNABEL

I don't want you to think I'm a pessimist, but I'm not as optimistic as you are.

GRANDPA

No, you never have been.

ANNABEL

It just feels like our compass is off right now. Like that story in the newspaper yesterday about the whole Northern Gateway Pipeline –

GRANDPA

Let's not go back over that again, darling –

ANNABEL

But just ... hear me out. Just ... in light of ... I don't want to debate –

GRANDPA

No?

ANNABEL

Just – British Columbia has one of the most pristine and beautiful coastlines in the world, right?

GRANDPA

Right.

ANNABEL

And we know that we all need oil to live.

GRANDPA

We do.

ANNABEL

And right now Canada needs to sell it to another country in order to maintain or even elevate our economic status.

GRANDPA

Yes, "but"?

ANNABEL

But we also know from past experience that you can never build a perfect oil tanker –

GRANDPA

Accidents will happen –

ANNABEL

Accidents will happen. And so my question is: What are we prepared to lose? Because no matter how wealthy you are, if you don't have certain things, your health, your family, and – in British Columbia, the concept of health is very directly connected to their environment.

GRANDPA

It is.

ANNABEL

So do we have to build a pipeline, sell to Asia, get rich in the short term, have an oil spill, deal with the consequences – Do we really need to go through that series of actions before we *learn* that over the long term what's valuable about British Columbia is its natural splendour?

GRANDPA

But, again, is Northern Gateway a higher-risk prospect than shipping the oil to eastern Canada and shipping it from there?

ANNABEL

But even if we could go through all of that analysis, Dad –

GRANDPA

I mean, right now if you look at the growth being created by oil being transported by train –

ANNABEL

But that's not my question, Dad. What I'm saying is, economically, will that province, will our country, not survive to tell the tale of a different kind of decision? Will we ... will we suffer so much economically?

GRANDPA

By not building it?

ANNABEL

Yes.

GRANDPA

No, I mean –

ANNABEL

Because –

GRANDPA

I don't know exactly what the economic impact is of building or not building. I'm not an oil and gas expert, darling. I mean, it will certainly have some push down on economic –

ANNABEL

But I'm not sure we even ask the question anymore.

GRANDPA

But look at all of the discussions we had twenty years ago about the North American Free Trade Agreement. And all of the naysayers there. But almost all economists will agree – and the one we've just done with Europe I think will probably be a pretty good one as well. So I think that Harper's doing some good things for the long-term economic well-being of Canada.

ANNABEL

This isn't about *Harper.*

GRANDPA

I mean, I'm not an economist but I don't think I've ever met an economist who would argue that free trade is –

ANNABEL

But what if you talk with someone other than an economist?! If you talk with doctors today, Dad, half their patients are depressed. And if you talk with psychologists today, 75 percent of their patients are divorced. And you talk with a child psychologist, 40 percent of kids are obese and are on drugs. Again, these –

GRANDPA

But these are people who specialize in those issues, darling. They wouldn't have a practice if they had no one to diagnose.

ANNABEL

But Dad, even *Ella* this year! (*pause*) There were a couple of moments when her teachers thought she was having attention problems. And when I told them that I absolutely did not believe in medicating children to fix academic performance –

GRANDPA

They wanted Ella to go on ...?

ANNABEL

Of course they did! They think I'm a bad parent. (*holding back tears*) Because I wouldn't fix the problem. But again, I'm talking about economic values leaking into places where it's not ... I mean, science, art, *education*, Dad. I mean, since when did performance become the ultimate ...? I thought we learned to understand.

ELLA enters.

ANNABEL

I didn't think we learned in order to perform. When your child doesn't understand something, you have an opportunity to learn together. "What don't you understand?" as opposed to "WHY DON'T YOU UNDERSTAND?" (*referring to ELLA*). This child is *ten years old*. It's just a really weird ... it's just a really weird compass we're using.

Pause.

GRANNY enters.

GRANNY

My goodness, is this all going into *The Watershed*? I mean, it doesn't sound like it was on-topic a lot of the time.

GRANDPA

No, no, we're all over the map here.

SCENE 17

BEATRICE enters and walks to the edge of the dock.

BEATRICE

Look at me, Grandpa!

BEATRICE jumps into the lake. ALEX enters and jumps into the lake. ANNABEL and ELLA join them. The actors playing GRANNY and GRANDPA watch the family swimming and transform into NEWS ANCHOR and JANICE DEAN.

NEWS ANCHOR

What's going on, J.D.?

JANICE DEAN

John, if this track comes true, this is the worst-case scenario what we're seeing here. Again, the worst-case scenario happening right now. I wish I were standing here making this stuff up. But here it is. Here's what I'm seeing. I'm warning you – *it's not pretty folks* – but here's what I'm seeing ... here's what I'm seeing.

END OF PLAY

ACKNOWLEDGMENTS

The process of writing *The Watershed* was long and circuitous, but one person triggered its creation at one fortuitous moment for one particular event. Thank you to Don Shipley, creative director of the Arts Festival of the 2015 Toronto Pan Am and Parapan Am Games, for trusting me with the commission that gave birth to this play.

Thank you to Chris Abraham for bringing my work to Don Shipley's attention at the right moment, and for working alongside me for more than three years to contain the dramaturgical heart of this beast.

Thank you to the actors whose talent and passion helped me refine the words on the page: Alex Ivanovici, Tanja Jacobs, and Eric Peterson, who participated from the start in preliminary script workshops, and to Bruce Dinsmore, Tara Nicodemo, Ngozi Paul, Amelia Sargisson, and Kristen Thomson, who joined us for final workshops and production rehearsals. Thank you to Julie Fox, Thomas Ryder-Payne, Kimberly Purtell, Denyse Karn, Merissa Tordjman, Angeline St. Amour, Oz Weaver, Monica Esteves, Normand Vincent, Jean-François Garneau, Kristina McNamee, and Lesley Bramhill, for making it possible to bring this new play to life onstage so boldly and professionally. Thank you to Joël Richard at Porte Parole for reminding me why it is worth producing ambitious plays.

Thank you to Andrew Kushnir, whose keen dramaturgical ear allowed me to detect my own voice among all the other voices. Thank you to Elle Thoni, who reminded me from the beginning to keep listening to the character of water.

Thank you to water for reflecting me back at myself so candidly.

The Watershed script development benefitted from a grant from the Canada Council for the Arts, and from funding from The Collaborations, an initiative of the National Arts Centre of Canada (NAC), English Theatre. Thank you to Sarah Stanley and Jillian Keiley of the NAC for their dedication to new play development in Canada. Early drafts of the play received dramaturgical workshop support from Playwright's Workshop Montreal (PWM). Thank you to Emma Tibaldo of PWM for her insight and encouragement.

Thank you to the team at Talonbooks, especially Ann-Marie Metten, for their commitment to this publication. Thank you to Merissa Tordjman for her meticulous work to make sure this version accurately reflected the 2015 production text.

Thank you to all the real people who took the time to speak with me so candidly on the record and to become characters in this play. Special thanks to Ella and Beatrice Ivanovici and Hazel Repo-Martell for their generous research collaboration and for reminding me that work can sometimes get done through play; and to Alex Ivanovici, who patiently accepted his wife's narrow portrait of his broad and beautiful self.

Thank you to my mother, Helgi Soutar, for teaching me about the value of a family legacy.

Thank you to my father, Ian Soutar, for showing me that one often has to swim upstream for answers, from mouth back to source, along a winding river of open-minded questions.

Annabel Soutar is a Montreal-based playwright and theatre producer. In 2000 she co-founded the theatre company Porte Parole Productions with actor Alex Ivanovici and she has acted as artistic director of the company since its inception. Soutar takes a documentary approach to theatre creation and since 1998 she has applied this approach to writing her original plays, *Novembre, 2000 Questions, Santé!, Seeds, Import/Export, Sexy béton*, and *Fredy*.

In 2012 *Seeds* was published in both English (Talonbooks) and French (Les Éditions Écosociété) and since then it has been produced in over a dozen theatres across Canada, including at the National Arts Centre of Canada in 2014, in a production directed by Chris Abraham.

In December 2015, the *Globe and Mail* named Soutar to its list of Artists of the Year.

Soutar lives in Montreal with Ivanovici and their two daughters Ella and Beatrice.